Yahweh of the Cosmos
Ultimate Symmetry in the Universe

by
Cedric Michael Guss

authorHOUSE™

1663 LIBERTY DRIVE, SUITE 200
BLOOMINGTON, INDIANA 47403
(800) 839-8640
WWW.AUTHORHOUSE.COM

First published by AuthorHouse 11/04/05

ISBN: 1-4259-0121-2 (e)
ISBN: 1-4208-9421-8 (sc)

Library of Congress Control Number: 2005909458

Printed in the United States of America
Bloomington, Indiana

This book is printed on acid-free paper.

Table of Contents

Did you hear the one about God and the universe?

I. Introduction

Anyone who has ever been interested in the question, who or what is God, will gain insight from reading this book. But we don't stop there. The how and the why are also discussed by focusing on clues left behind by the creator of his creation. As a natural consequence, this book focuses on the known scientific facts of the universe versus what we don't know, along with theories of creation, and it dares to tackle fundamental theological questions about the nature of God and our relationship to him. Many books by scientists often focus on "the what" of our universe because that is what they know. Albert Einstein once stated "The most incomprehensible thing about the universe is that it is comprehensible." Of course, that's easy for someone like Einstein to say; but what about the rest of us? This book tries to examine "the what" from a layman's perspective; and then goes on to speculate somewhat on "the why", something most scientists will treat like a hot potato, as in don't touch it. Those who enjoy not only science, but theology, philosophy, or cosmology will take a keen interest in the ideas that follow, as we explore the topic of creation along with identity and personality of God, and our place and relationship to him in the universe, from two very distinct cultures of thought – science and religion.

This book is not intended however to be an academic study of cosmology, theology, or science however. It is rather meant to give the reader a philosophical approach toward God and our universe based on what we know. It is presented in a way that people with diverse backgrounds can appreciate the concepts and speculations that are laid out, and should be understandable to people of any background or interest. People interested in philosophy will be particularly interested because we discuss the implications of recent discoveries that impact our knowledge about how we got here, their meaning for the universe as a whole, and the nature of a God that could have done this. Issues from both scientific and religious perspectives are explored and common ground is sought wherever possible.

Postulates in this book that relate to God or the origins of the universe are formulated by using a mix of common sense, logic, and available scientific information and theory, given what we already know about the origins of the cosmos versus what we don't know; some of the ideas presented here are profound, and others more obvious. The reader can decide. Postulates, where they occur, are not the major focus of the book however; they are the author's ideas and not meant to be misunderstood as widely accepted theological or scientific thought. They have not been proven or unproven, and the reader can choose to agree or disagree. But the real aim here is not to prove a new theory, but rather to enhance and enlighten the debate about God and our own universe based on what is now known from various circles of research in theology, archeology, cosmology, and yes, basic experimental and theoretical physics.

Speculative and interesting models that extrapolate on known facts are also presented that try to explain things, such as why extraterrestrial intelligence hasn't been found, which led, for instance, to the infamous question of the legendary physicist Enrico Fermi, "Where is everybody?", or what caused the Big Bang and breakdown of perfect symmetry that gave rise to our universe to begin with. Geometries of the universe are also discussed that try to explain cosmological problems that still exist among theorists and philosophers alike today; you could say that maybe things are wrapped up in a way nobody ever expected.

As an author, I write this book to present my own unique perspective, brought up as a practicing Christian, yet trained as a doctoral scientist in experimental high-energy physics, with several years of post-doctoral research experience; and thus while retaining my own insight into both religion and science, any personal bias toward one view over the other (where they could still disagree) hopefully is removed. Part of the fun in writing or reading about topics that are in this book is in the learning and discovering of various biases still exist today in the scientific or theological communities, and then trying to account for them. As we continue to learn more about our universe through scientific observation and theological deliberation, my guess is that scientists and theologians, as they begin to put away those biases, will one day find out that they are speaking much of the same language. And that is what we hope to commence here.

When someone asks me how to picture God, I often respond that a doctoral thesis advisor as seen by his poor graduate student is a good starting reference point. He knows how to strike the fear of God into you, he knows how to make or break your future, and he's often smarter and certainly more experienced than you are. No better was that made clear to me than the night that I realized for the first time whether I was truly going to finish my thesis. It was 1987, and we were preparing to show preliminary data from an initial test run of my thesis experiment to the directorate at Fermi National Laboratory (Illinois) the very next day. One set of detectors in that data had an almost dead zone that made it difficult to see the signal we were looking for, although another set of detectors was working fine. If we could present agreeable results from both sets of detectors, it would be much more likely that our collaboration would be granted another full year of data taking from the laboratory, so that I could finish my thesis with another run of the experiment. Little did I know at the time, as a naïve graduate student, how important this decision of the directorate would turn out to be, or that the Director of the laboratory, Leon Lederman, who would make the final decision, would be the following year's Nobel Prize winner in physics for a neutrino discovery that he made many years earlier. It is no small understatement that the pressure from my advisor to get some result that night was making me hypnotic; it seemed that the weight of the whole collaboration was in my hands. What we finally presented the next day was one data set where both sets of detectors were found to be working; and it was by pure luck (or that prayer to God), that I had accidentally stumbled upon that one data tape the previous night. In 1988, we were granted that extra year of data taking, Leon got his Nobel Prize, and I began writing my thesis. His Nobel Prize was a pure coincidence, you understand!

Of course, another person's picture of God can be quite different, especially for children. For instance, the story is told of a small town pastor who enjoyed making parish visits around town on foot; but he would take the dog along for a walk at the same time. One day he came to visit one house with children at home who happened to be peering at the window to see who was coming. When they saw him coming up the driveway with his dark suit, collar, cross, and the dog in tow, the children ran after their father shouting "Daddy, daddy, here comes God walking his dog!" As the pastor was not exactly an imposing figure, those children's view of God was quite modest indeed! Now, another child might have a more intimidating image of God, or one of a graying old man with a beard hanging down to the floor. And as we mature and get older, even our view of God comes under constant scrutiny and change.

For theologians and scientists, visions of God can get quite bizarre, without a doubt. And of course there are those, especially in these professions as it turns out, who simply get pleasure at trying to disprove the existence of God all together. While this book is a serious work, one hopes the reader will find the humor in all the ongoing sideshow wars of opinion, feeling, friction, attitude, and history between science and theology that are touched upon. Not only are the relationships of cosmology, physics, and theology considered in regard to the origins of the universe, but so are many facets and characters that led up to the various doctrines, theories, and ideas that are behind them; in doing so, the clues left behind in our universe, and the people who discovered them, can help us construct a fuller image of God than anything we might have learned from Sunday School. One hopes the reader will gain insight and find appreciation for these subjects in order to find his or her own beliefs, or lack thereof, in relationship with God.

It turns out, quite coincidentally, that in the first year of this book's availability (2006), we come across the 50th anniversary of one of physics' most fundamental discoveries, the lack of quantum spatial symmetry in our universe. And the person who first suggested it was none other than the mentor of my own trivial image of God, my eventual doctoral thesis advisor. Its actual discovery and interpretation led to a Nobel Prize for a set of Chinese physicists Tsung-Dao Lee and Chen Ning Yang in 1957. As we shall see, the significance of this discovery led to the tumbling of other symmetry laws of nature that have helped to solve one crucial riddle of creation: why we are all here anyway, existing as a bundle of interacting particles of mass without its set of corresponding antiparticles that would have annihilated us into nothing but a beam of light in the first place. And given why we are all here, how did it happen; we're only now beginning to understand the answer to this question also.

And of course that leads to the proverbial "who dunn it", and why did he decide to do it. Extracting clues left behind from the universe's origins, given what we have learned from experiments in physics today, have helped to reveal secrets of our creation that have been locked up throughout the ages, and will help to answer these questions also as we go along. We will see that these secrets have everything to do with the loss of symmetry in the universe, and what will be needed to restore this symmetry. And when we talk about the symmetry of anything, there will naturally be its most fundamental feature, its center; and that center may well lead us to God. Undoubtedly we are all probably going to be surprised, as he himself must be full of surprises. For instance, he is not likely to be what I call that three-headed monster so often described as a Trinity by Christian doctrine. We'll see how that came to be, as well as how we can put the pieces of the puzzle together from what we don't understand in the Bible to what we do understand about our universe today. The problem of course is that the symmetry of the universe is now missing; and where symmetry is lost, so goes its center. If we are ever going to find God in our universe, we will have to look for that lost symmetry as well; and the clues left behind from creation may well provide an answer.

II. Outside the Box

When we think of God, we seem to have a human need to personalize him in a way we can understand. We think he has a mind, arms, and legs like anybody else. Indeed, the seeds for this thinking may have been laid in the biblical Old Testament which says we were created in the "image" of God. Well, this is all fine and good; but the problem is that it leads to questions like where did God come from, where is he going, what is he made of, is he male or female, and so forth. If we are ever going to be able to understand God and get a handle on who or what God really is, we are going to have to begin to understand him as being outside the box of our creation. For instance, if he created space and time, then he must be outside space and time. And if he is outside space and time, then he must have no beginning and no ending. His "image" is not set with human boundaries or components or features. He must be something else. It is this something that we strive to understand from the clues he left behind during the creation of his cosmos.

Other common problems that naturally pop up are issues regarding the seven day creation story in Genesis, evolution of mankind, resurrection of the dead, extraterrestrial life, and construct of the Christian Trinity. These are all issues subject to human history, culture, translation, and initial origin of writing. Some say the Bible must be taken literally, others say it must be taken figuratively, and still others say it is some combination with the culture of the times taken into account. Well again, all these issues can be better understood from the perspective of where God actually exists, not only inside, but also outside that box of space and time called our universe.

For instance, there is that lovely creation story so beautifully written in Genesis which describes God's creation of the universe in just six days, while he rested on the seventh day (as if he needed it!). It states that the wind of God filled the void, expanded the heavens, and creation was formed from a chaotic emptiness. Many times, these events or illustrations can be interpreted physically, and other times only figuratively. When the Bible says that God "stretched out the heavens", that fits well with the cosmological theory of inflation in the early universe, which we'll discuss. The chaotic emptiness can easily be seen as a description of quantum mechanics, which we'll also discuss; we'll see that this becomes one of God's most useful tools for getting the Big Bang off the ground in the first place. And then there is that wind, or spirit, if you will. For people of faith, it is a comforting notion that God fills the universe with himself; but it begs the question how did he come to do that? Questions like this are easier to answer when we think of God outside the box; then we can pursue his interface to our existence. And this interface is one of the things we discuss later on.

And then what about that guy who wrote the story about how this all occurred. Judeo-Christian teaching has historically suggested that Moses wrote the first five books of the Old Testament: Genesis, Exodus, Leviticus, Numbers and Deuteronomy. Only problem is: there's an account of Moses' death in Chapter 34 of Deuteronomy. Now that's interesting; perhaps I should write my own obituary also. Someone already published that I was dead; so I might as well write my own! Oh well, I'm always late for everything anyway; so I might as well be late for my own funeral too!

Now don't get me wrong; I don't wish to denigrate fundamentalists who believe in the literal account of the Bible. On the contrary, the literal nature of what is written there can be preserved at some level by just trying to understand both the writer and God a little better;

and our observations and scientific discovery in the cosmos can help us do that. For instance, what was one day to God at the time of creation? He created the timeline of day to night and so forth; but it took a long time for the earth to form (the universe is at least 13.7 billion years old, and earth 4.5 billion years old). And let's not forget that the earth had to rotate on its axis in order for our concept of a day to occur, almost like God spinning his top, as it were; and even when it started spinning, the earth spun so fast that a day started out being about 6 hours or so. Now St. Peter made an interesting point in his second book of the biblical New Testament, that one day is as a thousand years to God, and a thousand years as one day; okay, maybe he was off by a tad. But what is probably going on here is that God's concept of a day is all together much different from yours or mine, if only we would allow our arrogant human nature to consider that God is outside of our little box called the universe!

In another example, there are those waters that God moved across, referred to in the Genesis' creation story, and its separation by the dome called the sky. God's concept of waters at the time of creation was probably not quite our own, if we just allow ourselves to think about it for a moment. Matter in the form of atoms took thousands of years after creation to cool and bond, much less the H_2O molecule that makes up water, although it is true that recently accepted geological thought has that maybe was indeed around on ancient earth, but at about 100 million years of age or so. However, just as is the case with the "image" of God, one can gain an understanding about what God was actually talking about with the writer of Genesis. There would have been a bobbling sea of virtual particles in the void of creation that existed from the start (and would be responsible for the heat energy in the big bang itself). Perhaps Genesis was just simply trying to suggest that God controlled this movement of the virtual void that would give rise to the energy and matter formation of creation itself. You also get the idea that the concept of gravity has to work its way in here somewhere, not that the writer of Genesis really knew what gravity was (not that many people do today either, come to think of it). But God would have known what it was, and maybe the dome was the best way for him to put it to someone. But that is not such a bad idea, when you consider how the warping of space forms around star or planetary matter, such as the sun or earth; the warping of space itself gives rise to gravity and has the same basic symmetric geometry of a dome. And thus, we can go on and on with this kind of description of figures in the Bible.

So okay, science has a lot to contribute to this discussion. We can't have a study of theology without guidance from what is known about ourselves and our universe. The story of creation makes that point clear enough. We will see that many famous scientists have grappled with the concept of God and how he put everything together; indeed it's what motivates many a scientist. Albert Einstein used to say "Science without religion is lame. Religion without science is blind." On that point, I could not agree more. If science is ever going to find the ultimate Theory of Everything, like some physicists and specifically quantum gravity theorists desperately try to do today, it will have to bring God into this equation somehow; and believe me, there is nothing a scientist hates to do more than that. Similarly, theologians had better get used to reading the science magazines if they ever hope to align religious doctrine with truth; we'll see that Galileo certainly brought that point home sure enough. They would do well to listen to St. Paul, in his New Testament letter to the Romans, when he stated "Since the creation of the world, God's invisible qualities—His eternal power and divine nature—have been clearly seen, being understood from what has been made." No one could have put it better. One of the things we hope to do here in this discussion is to gain a greater insight of the creator by observing and analyzing what he has created.

Unfortunately, religion sometimes get sidetracked by taking stands on issues that it really knows nothing about, whether it be the creation of the universe or the supposed geocentricity of the earth or the concept of the Trinity itself. On this latter issue, for instance, my guess is that most Christians don't really know what the Trinity is, except that someone in a church somewhere must have told them that it was so. They certainly won't find anything in the Bible about it, nothing by the prophets about it, nothing by the apostles about it, nothing in the writings of St. Paul or St. Peter about it, and indeed nothing by Jesus himself about it. With all of Paul's brilliance and with all of Peter's personal witness, there was never once a description or discussion of God as a Trinity; in fact, they never used the word "Trinity" at all. And if the Trinity is so important, then somebody forgot to mention it to Jesus. Yes, they all brought up the concepts of Father, Son, and Spirit, which are three prominent revelations of God written throughout the Bible; but they understood better than anyone that they were dealing with someone and something that was much greater and grander than that which could be described by simply three human characterizations of God, namely someone who could be contained by our little box of space and time and in our little corner of creation. We'll see that this concept of God as Trinity was really a fourth century concoction by some learned men of Greek philosophical persuasion who made a hopelessly naive human attempt to understand the creator and his creation within the framework of human identities and relationships, as if he belonged to our silly little world. My father, an ordained Pastor of some 50 years plus within the Central Pennsylvania Synod of the Evangelical Lutheran Church in America, once gave a good description of the Trinity as like watching three windows in a room that look out at the same sky. This is very useful; but I always have a nagging inquiry: why is God described as a discreet mathematical quantity? From the perspective of the physical universe, we'll see that these windows can get quite small and can get quite large, and all at the same time. And it might be hard to imagine just how small and how big they could become.

As residents of our little box called the universe, we are sometimes forced to take what we know that is inside the box, and extrapolate to that which is outside to gain some measure of the true nature of the realm of God. But in order to do this, the first thing we have to do is to free God from this box of universal space and time, since he is the one that created the box in the first place. Then we will see that questions about how God got here and where he is going, and indeed how we got here and where are we going, will be a lot easier to answer. It would also help to give scientists a better perspective on the physical origins of the universe, because they are always much more quick to find the what and how than they are swift to answer the who and why; meanwhile, theologians can typically do the reverse. When and if the two sides ever get together, there just might be some real progress in our final understanding of God and the cosmos.

But that brings us next to the historical friction that has existed throughout time between church and science; and what has happened to get us beyond it. The two, not surprisingly, did not get along very well in the past, but surprisingly get along much better in the present. We will see what that was all about, and where the suspicions still surface; more often than not, the arrogance of the players on each side has played a decisive role. From there, we'll get on with the story of our universe, the mind of someone who would have created such a thing, the problems that are still not understood, and what they allow us to say about the plan of God, and where we seem to be headed. And we'll see that science and cosmology may have stumbled upon more than they were originally bargaining for, the consciousness of the universe. And as the cosmos expands seemingly ad infinitum, things may well get wrapped up together more than anyone might have expected.

III. Church and Science

Perhaps there is not a longer cultural war than still exists between church and science. The two find themselves in this predicament because of the methods of thought used by each culture. Scientists seek to find truth through observation, measurement, and analysis. Extrapolation and interpolation techniques are only employed when the target measurement remains hidden by other more significant factors or events that only allow a pure measurement at a nearby point. Scientific truths are always subject to multiple and repetitive measurements with other techniques as more advances are made in the tools of measurement. Theory tries to explain things that have not been measured, based on things that have been measured, and is always subject to query, further investigation, further measurements, and improvement of ideas.

The church on the other hand relies on theology and doctrine that has been written down sometimes over many centuries by people who have become heros by their sacrifice, accomplishments, thoughts, and witness. Quite often, when doctrine is agreed upon and written down by one generation, it becomes quite unmovable for future generations to make changes. It is never subject to active query or update or analysis because of its rigidity. The usual explanation for this approach is that the truth upon which doctrine is based is not subject to or dependent on the culture of the day. As we shall see, this approach has led to several mistruths put forward over many centuries, most obviously being the church doctrine of the geocentricity of the earth's position as a center point in the heavens. Unfortunately, the church has been guilty all through its history for ignoring science, advancement of thought, measurement, observation, and reason in order to maintain its doctrinal status quo, whatever that may be.

Therefore, as science advances and more observations and data become available about our existence, the culture of the church naturally lends itself to evasion, avoidance, or ignorance of the facts as they are received by science for the mere reason they may contravene an existing doctrine within the church that is taken as religious truth. When this phenomenon occurs, scientists and church leaders find themselves at odds with one other; and the result is that one refuses to listen to the other, or worse, one demeans the other with arrogance and self-righteousness. Then, those immortal words of Einstein which imply that science and religion need each other come to mind. If we are ever to have true advancement of thought in our world, science and religion simply must learn to understand and accept each other, and use what each other has to offer to advance their progress. At that point, we will finally have a chance to gain a final understanding, not only of creation and how it happened, but also of its creator and why it happened.

Galileo (1564-1642) may be the most famous scientist of all time that got into trouble with the church. It seems he got into hot water with the Roman Catholic church when he refused to agree with the geocentric Christian doctrine that the earth was the center of the universe, even though his telescopes were telling him otherwise. It isn't so much that the church was so opposed to the idea that the earth was not the center of the universe that landed Galileo in jail, but that he insisted that it was in fact theological truth. Well, theologians would have none of this from a layman; so Galileo in the end was forced to recant. In fairness to the church, Galileo also thought that the sun was the center of the universe, which of course it is not; at least the church could be credited for not making a rash acceptance of new science. But give credit where credit is due. Legend has it that Galileo was the first to measure the actual acceleration of earth's gravity (32 meters/second^2), by hauling a cannonball and a rock to the top of the Leaning Tower of Pisa and then dropping them at the same time from the roof in order to show they had the same gravitational acceleration. More likely though, he was trying to demonstrate the effect by rolling weights down a hill to everybody's bemusement. Either way, he learned that by dropping items of different mass to the ground at the same time, he could demonstrate that gravity was the same on any mass, regardless of its size.

Copernicus (1473-1543), a century earlier, had made similar observations about the earth rotating around the sun; however he was a little more spry about how to handle the church. When he published his theory of earth's solar orbit, he dedicated the work to Pope Paul III. Needless to say, he had a better time with the church. Unfortunately, he also gave his work to a Lutheran clergyman named Andreas Osiander (1498-1552) who in turn gave it to the great church reformer Martin Luther (1483-1546). That was a mistake, because Luther trashed the theory; and hence it would end up taking Protestants a much longer time to embrace it. For whatever reason, the church wanted to stick with the notion that the earth must be the center of things, a theory they had picked up from the great Greek philosopher Aristotle (384-322 B.C.) and later "confirmed" by an Egyptian mathematician named Ptolemy (100-170 A.D.), who seemed to show that other planets, including the sun, were spinning about the earth, and hence, so the church thought, were heaven and hell! It's an idea that lasted as official church doctrine for 1600 years! Perhaps the idea was to scare everyone out of their wits, with heaven and hell looking at us from overhead! And Luther, always one for worrying about demons, the devil, hell, and the like, would not have liked the idea that there was no place for them near the earth!

Then there is the case of Giordano Bruno (1548-1600). Bruno was a bit of an odd duck, who studied to be a Dominican monk, dropped out, and then became a philosopher and a magician all at the same time. He often went around Europe looking for an audience who would listen to his lessons on philosophy and at the same time watch his magic show. He was living at a time when science had started in earnest to break away from philosophy and religion, and was just one of the few who agreed with Copernicus' view of the solar system. In fact, he thought Copernicus' theory introduced a new celestial instability to the universe. He reasoned that if earth really wasn't at the center of the universe, then there must be a plurality of worlds that exist, hence multiple crucifixions of Jesus, and multiple resurrections; and I'm not kidding. As I said, he was a bit of an odd duck; but he probably was just trying to mock Christian doctrine. He thought Christianity was irrational because it disagreed with other faiths and was entirely based on faith with no scientific basis. That view would probably have many takers today. He also wrote that we live in an infinite universe (the first to make such a claim) that left no room for God (if he only knew!). He saw God and nature as being one and the same, which is opposed to most worldwide religious views of God, but strangely close to pagan Greek philosophy, which has given us, quite ironically as it turns out, the concept of the Holy Trinity.

Well, you can imagine what the Catholic church had to say about Mr. Bruno. After he had returned to Italy, he was snared by Venice authorities who turned him over to the Pope for imprisonment. After seven years and still refusing to repent, Bruno was blindfolded, gagged, and burned alive at the Campo di Fiori, (The Field of Flowers) in Rome near Vatican Square for his heresies. He exclaimed in a letter near the time of his execution that "the Lutherans seemed more civilized than either the Catholics, Calvinists or Puritans". He didn't explain, but perhaps he thought the Lutherans were the most willing to have open minds about his ideas; too bad they didn't feel that way about Copernicus though. Martin Luther had nothing to do with Copernicus in his time; so I doubt he would have had much patience for Bruno either. But the combination of the two might have made for delicious conversation, because Luther had superstitious tendencies and believed demons were everywhere, while Bruno treated superstition with derision and saw only stupidity where there was any talk about spirits. However, starting from the 18th century onward, European philosophers have once again picked up on some of Bruno's work, and have realized that he wasn't as much the quack that he seemed on first blush. Actually, he was quite intelligent, maybe too much so. Long before Einstein, he talked about the relativity and infinity of space, eternity of time, implications and reality of endless change, and conservation laws of physics. In many ways he instituted his own science of cosmology. But scholars also acknowledge that he had a streak of uncontrollable and arrogant independence which probably brought about his bitter end. He once actually met with Queen Elizabeth and later wrote that she was a diva! You might have been able to guess that she didn't think much of him either.

Then of course, there is the greatest scientist of his time (and perhaps some would say of all time), Sir Isaac Newton (1642-1727), who gave us our classical universal laws of gravitation that are useful even today, and born the same year that Galileo died. He was a deeply religious man. Yet he became disillusioned with Christianity because he believed it had strayed from the original teachings of Christ. He was particularly upset with the concept of the Trinity, and how it came about. We'll discuss more about the Trinity later; but Newton rejected it on principle because of its lack of a scriptural basis. However, the Church of England required him to take holy orders in order for him to accede the Lucasian chair, as professor at the British university named, you guessed it, Trinity College. (The present Lucasian professor today in Cambridge is none other than Stephen Hawking.) Luckily for Newton, the Church of England wasn't as strict about adherence to doctrine as the Catholic church; so King Charles II issued a decree that he didn't have to take the holy orders after all, in order that he could hold a chair at the university. Of course, it might have also had something to do with the fact that Newton was considered to be one of the most brilliant people to have ever come along. But Newton decided to keep his views about the Trinity private anyway, so he wrote a private treatise about the Trinity, among other frauds he felt had corrupted scripture during the 4th century, to a philosopher friend, John Locke (1632-1704), for the purpose of having them published, but only in France. Good thing too, his successor to the Lucasian chair, William Whiston (1667-1752), was in fact ejected from the chair by the Church of England for his rejection of the Trinity; and he was even a Protestant minister's son! Seems familiar! By the way, as we get into our time machines and reach the 24th century, the Lucasian chair falls to none other than

Lieutenant Commander Data in the final series episode of ***Star Trek – the Next Generation***!

The struggle between scientific thought and religious rigidity took a new turn in the 18th century during the age of Enlightenment, as scientific study and knowledge was being encouraged for everyone. Except for that bit about the angels still being responsible for the pushing of the planets around the sun (even after people accepted the idea that the earth wasn't at the center of things), people in western civilization were more likely to question authority, religion, and politics in their countries as well as the church. Some of the consequences of this age, of course, were the revolutions in America and France, as majestic and magical royalties were being rejected. The church gave heed to the idea that scientific study and knowledge was in the interest of humanity. Society finally gave heed to the very Roman legacy that it could live without religious supervision, but not necessarily without religion. The idea that mankind was doomed was a Medieval notion that was replaced with the idea that there was hope for mankind after all. Scientific research and study flourished from then on. For one thing, government funding for research increased, and has steadily done so to the present day. Of course there were hiccups along the way. For instance, a fellow named Charles Babbage (1791-1871) had built the world's first computer in England in 1840 using available gadgets and switches available in the day; and yes, he also had held the Lucasian chair at Cambridge. Unfortunately, when he appealed to Parliament for more funding to build a better one, the English Prime Minister Robert Peel (1788-1850) turned him down citing its uselessness, unless of course, as he said, it could keep time! Scientists despaired; but at least at this point, there was no religious controversy involved.

In 1859 however, a real curveball was thrown at the church that it has never yet really recovered from. Charles Darwin (1809-1882) published his book ***On the Origin of Species by Means of Natural Selection, or The Preservation of Favoured Races in the Struggle for Life;*** then the gloves were off. In it, Darwin discusses how natural selection drives a species toward favorable variations which then get propagated to the next generation and so on, leading ultimately to new species being created over great expanses of time; but offspring would be produced faster than the food supply could be produced, thereby causing a struggle for existence and survival of the fittest. Although this work had shocked many churchmen, who believed that humans all started with Adam and Eve (in the biblical book of Genesis) frolicking around in the garden so that we didn't just happen from a couple of smart apes, it turned out that many scientists of the time were also churchmen willing to open their minds to new ideas as research became available; and those who were unwilling or unable to examine the scientific evidence for evolution just dismissed it out of hand. Even in modern times, you have organized public officials, such as the 1999 Kansas Board of Education in the United States, who have tried to shut down the study of evolution in public school. The National Academy of Sciences itself counters with statements such as "religion and science are separate realms" in public education and should thus be taught as such. Fortunately however, most government and church officials in western society now allow the study of evolutionary theories to be taught in parallel with the study of religion in the classroom. Unfortunately for Darwin though, he would have to wait another century before his discoveries would be finally recognized by religious leaders. Pope John Paul II (1920-2005) for instance even acknowledged that Darwin's work was more than just a theory, but actually a welcome advance of knowledge and consistent with the faith. We'll see later however where the theory of evolution may actually have a basic flaw; and it comes about from our understanding of modern physics, no less!

Then there are those scientists even today who laugh and scoff at the beliefs of some religious people who dare to take the biblical story of the creation of the universe seriously, because they purposely dismiss or ignore theories of evolution and the Big Bang with no basis to do so. Carl Sagan (1934-1996), an accomplished astronomer (of "billions and billions of stars" fame) and professor at Cornell University (a place we'll visit again in our quest of the creation) was quick to point out in his writings that the church's continued disagreement with evolutionary theory was "unproductive" and that organized religion had been a perpetrator of "deceptive actions".

However, he does go on to say that religion in its purest form, e.g. without its authoritarian bureaucracy, has a mystic core that seeks to appreciate the majesty and beauty of our universe. But again, he associates many things that religion claims are mysterious as the duty of science to explain and understand. Well, this kind of thinking is all very nice, and many scientists no doubt share this view, but it can be argued that it is also an arrogant way of thinking. This kind of arrogance is as much a failing of science as denial and avoidance are of religion. If science ever hopes to find the final Theory of Everything, it is going to end up vastly disappointed if it doesn't find time to consult religion. One clear example of this of course, and it would have been right down Carl Sagan's alley, is the understanding and rationalization of the realm from which our universe was initiated, i.e. where did the material, energy, and heat for the singularity that caused the Big Bang come from. Science can never win by answering this question on its own, because the realm that created it would have been outside the parameters that science knows, or ever hope to know, how to measure. If at some point the arrogance of science doesn't give heed to theology or philosophy at the edge of its known physical boundary, then it will lead off into a tangential direction that is hopelessly void and wrong. Of course, one can speculate, as Stephen Hawking does, about that nature of that physical boundary, and whether it exists in the first place; but even he acknowledges that God may well be behind that "iron curtain" of our space and time. And thus, there just isn't anything we can do about it.

Similarly though, religion has to be willing to accept those foundations of knowledge that science has been within its bounds to observe, measure, analyze, and understand; when the arrogance surrounding the doctrine of religion refuses to do this (as happened with Galileo), then it too becomes hopelessly lost and void. For instance, some Catholic theologians apparently still have some problem with the study of the moment of creation, and how it must have all started. But it could also be argued, as it could be done with the anti-Darwinists, that worrying about the study of how God must have started everything is only an effect of comprehending how God exists inside our "little" box called the cosmos, where he seemingly can't be found. If God fits only inside the box, where is he, where did he come from, and how could he have created us anyway? It's like the old proverb, can God create a rock that's too big for even him to move? Once you manage to put God outside the box of our knowledge of space and time, the whole problem goes away, because it becomes much easier to see how these questions can be answered. One of the things we will do here is to try to understand God from clues left behind in our asymmetric universe, making the basic assumption that God must be part of the ultimate symmetry of his realm, and, by extraction, our universe as well; thus we can gain more insight about how God "did it" in order to create our universe, and where the bridge between him and us lies.

As for the initial creation of our universe, it must have worried even Pope John Paul II, because he once told a group of the physicists, "Any other teaching about the origin and makeup of the universe is alien to the intentions of the Bible, which does not wish to teach how heaven was made but how one goes to heaven." Among those present with that particular group of physicists was none other than one Stephen Hawking, renown theoretical physicist and Lucasian chair at Cambridge University. In his best-selling book *A Brief History of Time*, he states that he was there when the Pope told them it was okay to study the evolution of the universe after the Big Bang, "but we should not inquire into the Big Bang itself because that was the moment of Creation and therefore the work of God." And then he added: "I was glad then that he (the Pope) did not know the subject of the talk I had just given at the conference—the possibility that space-time was finite but had no boundary, which means that it had no beginning, no moment of Creation. I had no desire to share the fate of Galileo, with whom I feel a strong sense of identity, partly because of the coincidence of having been born exactly 300 years after his death!'" So certainly he didn't want to get the Galileo treatment, much less the Bruno one. Well, if it makes Hawking feel any better, Pope John Paul II had absolved Galileo of his "crimes"…not that it has done much for Galileo today, you understand.

Now the church as a whole has thankfully matured to a level where its leaders are not worried about the discoveries of scientific research. The church, to its own credit, has been able to stand on its own through thousands of years of bickering and hate mongering of people of all stripes, and it has learned since the Middle Ages that its theology, by and large, will stand up to any discovery by modern day science. In a funny twist, theology and science are coming more into convergence, albeit in an uncomfortable but interesting and amusing one at that. Pope John Paul II himself convened many gatherings of scientists at the Vatican during his pontificate, such as the one with Stephen Hawking, and embraced much of the research that has gone forward in the fields of evolution and cosmology.

I never felt a problem personally by being an active member within the church and being a research scientist at the same time, not that anyone really cared. But bishops, pastors, and congregants have always treated me with a great deal of respect as a student and researcher in science. Unfortunately, I wish I could say the same when the shoe was "on the other foot". You see, when I am in a crowd of scientists, I am sometimes sneered at for being religious. In fact, I remember one colleague who once had the gall to confront me directly about how I could be a legitimate student of science and actively believe in a spiritual creator of the universe; his thrust of course was that I could not conceptualize ideas and handle data appropriately because I had some belief in mysterious powers at work in my study and research. It's called the scientist's snob scale; and scientists could even learn a thing or two from religious leaders and theologians on the proper use of this scale. You see, some senior officials in the church are not immune to this behavior, even among themselves. In the Roman Catholic church, for example, there are a few priests who denounce politicians of their own faith by saying things like one is not saved and cannot take their sacrament of communion if one doesn't believe the way they do (like taking their stand on abortion or accepting their view on the substance of communion without being called a heretic). None of this nonsense ever solves anything, but to create division and bitterness, because sincere people in good faith just sometimes disagree, given each person's own experience and upbringing. The church's primary mission is to save souls, not reject them. Similarly, science's mission is to the truth, not to bias.

As it turns out however, some of the greatest scientists of all time thought about God quite often. We already mentioned Isaac Newton's travails with his faith. Albert Einstein (1879-1955), whose accomplishments include the Special and General Relativity Theories, advancement of Brownian motion, discovery of the photoelectric effect, for which he would receive a physics Nobel Prize, was quite possibly one of the two greatest physicists of all time (along with Newton). Einstein was brought up in a Jewish family and had a "deep religiosity" about him as a child. (Leave it to Einstein to describe his personal faith in some abstract way.) However, he became a skeptic as he grew up because some stories in the Bible just did not add up for him; too bad he never had the chance to read this book! Yet he never let go completely his concept of God. He often used God in his quotes, such as the time he blurted out this little gem: "I want to know His thoughts, the rest are details." Another time he complained about the newly developed study of quantum mechanics during his early career by having to drag God into the argument; "God does not play dice with the universe", he quips. Of course, Werner Heisenberg (1901-1976), the primary architect of quantum mechanics and the Heisenberg Uncertainty Principle for which he too would earn a Nobel Prize, had a good retort for Einstein; when asked what he would ask God if given the chance, legend has it that his response was in the form of simply two questions: "Why relativity? And why turbulence? I really believe he will have an answer for the first." (Is it just me, or do you get the feeling these two guys had at least a colloquial dislike for each other.) So here is Einstein, originally from a Jewish family fleeing Germany when Hitler took power in 1933; and there is Heisenberg, a leader in Nazi Germany's atomic bomb project during World War II, two towering leaders of modern physics, ribbing each other's theories almost semi-seriously from across the Atlantic, with God in the center providing the justification for each other's arguments!

Then there was Niels Bohr (1885-1962), who had come up with a modern understanding of the structure of atoms, and in particular hydrogen, for which he would also win a Nobel Prize, had choice words for both of these figures. With Einstein, he once retorted "Who are you to tell God what to do!"; and with Heisenberg, his most famous former student with whom they had become best friends, he began ignoring all together when he found out Heisenberg had been working on the Nazi atomic bomb project. As an aside, Bohr would later consider rejoining his relationship with Heisenberg after the war with several letters; but he never sent them. Scholarly debate exists to this day what became of their relationship, and a famous pact that Heisenberg said he made with Bohr which stated that scientists on either side should not to work on the atomic bomb, a pact Bohr later dismissed as a fabrication.

Quotes of God of course didn't stop with these leading scientists. There was another leading figure of quantum mechanics, Erwin Schrodinger, whose namesake was the most famous physics equation in the history of particle wave mechanics, the Schrodinger's equation, and who said that science has its limits because it "knows nothing of beautiful and ugly, good or bad, God and eternity. Science sometimes pretends to answer questions in these domains, but the answers are very often so silly that we are not inclined to take them seriously." Well, at least my postulates will find agreement with that statement! Then there was Richard Feynman (1918-1988), Nobel Prize winner and author of Quantum Electrodynamics (QED) who once said "Many scientists do believe in both science and God, the God of revelation, in a perfectly consistent way." And he is right; present American surveys have shown that while 44% of the general population goes to church on a typical Sunday, 43% of Ph.D. scientists do also. And I happen to be one of them (QED)! But then there was this quotable gem from "the avowed atheist" Feynman: "God is always invented to explain those things that you do not understand." It turns out Feynman had built his own theory about why religious students of science can turn into atheists: they just start learning how to doubt. And of course, there is Stephen Hawking; in his book *A Brief History of Time* he mentions how nice it would be to know why we and the universe exist because then we would know "the mind of God". And lastly, but not least, there is 1988 Nobel Laureate Leon Lederman who refers to the yet undiscovered Higgs Boson as "the God Particle" in his book by the same name, a theoretical particle that could well explain the makeup of universal mass itself, which we'll be talking about later.

Now some agnostics and atheists will say these scientists simply saw God as a mathematical construct of events; but it is probably a little more than that. Otherwise, there wouldn't be this fascination with God. Sometimes though, it might be the agnostics and atheists, who look to science for all the answers, who might have some explaining to do; the models all must start with something from somewhere. People like Schrodinger and Hawking have admitted that there may well be a realm where science may simply break down; and I might think they probably spent a little more time thinking about this realm than your average run-of-the-mill agnostic. Thus I offer a speculative postulate that there is a realm that exists from which our universe was born where the laws of the physical science of our cosmos do not apply, and therefore the rules governing the origin of the Big Bang itself will never be explainable from *within* the physical space and time boundaries of the cosmos. This does not mean, by the way, that other universes similar to ours do not exist. They could exist by simply being created in the same way as ours, or they could exist with another set of physical laws dissimilar to our own. The only common thread in all of this is that there is another realm beyond our universe that allowed the singularity of infinite density and energy to expand into the universe we call our own. And here is where perhaps some scientists make a terrible mistake. They naturally assume that the laws of physical science must apply to the creation event itself, i.e. the laws that allowed our singularity of existence to occur in the first place. That is likely to be an unwise assumption. Where there is an admitted bias in ideas and theories discussed here, however, is that the basic assumption and postulate that God himself exists naturally within his own realm caused or allowed our universe to form as a subset; and hence God climbs into and crawls out of the box called our universe as he chooses. The fascinating question therefore is how does God provide this interface between the box that is our universe and his realm of which we are a subset? From clues left behind in the creation of our universe, we can acquire ideas about the real answer to this question.

On the front pages of major newspapers you can often find articles about the culture of science versus religion. It's a popular subject for many people of various backgrounds; and it leads to some very strong emotions for some of them. However, when you have leading scientists being quoted as saying, you can't believe in God and be a serious scientist, or that one of the triumphs of science in the modern age has been to undermine religion; you begin to see just how serious and real the problem of arrogant bias by the scientific community itself against religion still exists today.

A Pastor of mine once told the story about how God once told a scientist how everything worked. When the scientist, wrapped up in all of his arrogance, heard it, he said "Fine, now go away; I don't need you anymore because I can do it all myself." God said okay, and let the scientist be. When the scientist later tried to create something by himself, he grabbed some dirt from the ground to bring into his lab. When God saw it, he said to him, "No you don't; go get your own dirt!" I suppose this story applies to even some prominent scientists today, or evolutionists who dream about how that glorious goo of amino acids formed into the first protein; but if you really do wish to believe in atheistic theories of creation, or how naturally we all evolved, someone somewhere still had to provide the heat, the real particles, the virtual particles, the atoms, the molecules, and, yes, even the vacuum energy, pressure, space, and time to have done it all in the first place. And no do-it-yourself arrogant wiz scientist is so quick with a decent theory about that, if for no other reason, science doesn't have one.

And as we go on, we'll see that our creation isn't simply about where we've been, but where we're going and what we can do (or not do) about it. Charles Darwin once grieved how badly it will be someday in the far distant future when man will have evolved into a much more advanced and excellent figure, only to see his universe tear itself apart from underneath him. On that point, Charles and I could not agree more. We can learn from the past to correct our own mistakes; but as for the universe, learning from the past will not change where we are going. However, just as we can learn from the clues of our own mistakes, we can also learn from the clues left behind by the universe to find just what is this realm of God all about anyway, and thus what we can do both scientifically and theologically to bridge that gap between us. And perhaps, this is just what God wanted from us in the first place.

IV. The Mind of God

I am tempted to begin here with the words "In the beginning…", but will leave those infamous words where they belong in the first chapter of the biblical book of Genesis. Besides, who am I to deal with a topic that I was not there to see. What I can do however is start with the mind of God, not because I have his, but because I have my own. First of all, you have to understand that God does not have a mind. We all want to believe he does because we have this internal devilish human need (don't ask me what it is) to see God as a human being in some way. But he is not. "I am that I am", as the burning bush said so eloquently to Moses. And this is the origin of the term Yahweh, which is Hebrew for "he is". This is how Jews originally learned to understand God and therefore to call him by name. After all, what else were they going to call him…Doofus, Ziggy, Homer, Frankenstein? Somehow I don't think so. It turns out Yahweh is the third person singular form of the ancient Hebrew verb, "haya," meaning "to be." It means a state of existence, or that he is self-existing. To understand this better, one needs to understand Jewish culture in the time of the great exodus of the Jews from Egypt, as described in the biblical book of Exodus. Here the writer (Moses by legend) claims this is how God refers to himself. Hebrew names in Jewish culture were sentences in and of themselves, and not simply names to distinguish two people apart. So it is that Moses refers to God by a descriptive sentence that is his name which simply means he exists. I also note that my use of the male pronoun in reference to God does not in any way imply God is male by any stretch of the imagination; it is used to maintain consistency with the Bible, and in particular with Moses. One can consider it for practical purposes to be nothing more than a convention of the language, whether it be Hebrew or English.

We see many books, written even by physicists, that are fascinated with the mind of God and what he is thinking and so forth; even the great ones like Einstein and Hawking talk about wanting to know God's thoughts or understand the mind of God. But I am here to set the record straight that God himself does not have a mind. Note carefully however that I did not say that God, as creator of our universe, is not **intelligent**. I am merely making the point that God is not a human being and therefore does not have human characteristics; but that does not suggest in any way, by the way, that he can't appear at any time as a human being if he wants to be. He can do that because he can. God isn't exactly for the faint-hearted!

Perhaps all this seems confusing, but it shouldn't be. A creator of anything can always appear as that which he/she has created without having to be that anything in the first place. Perhaps a better question is where does God obtain his intelligence? We know it must be there just from observing the organization of the universe from a state of pure chaos. And hence, we are witnesses to God's invisible qualities of eternal power and divine nature, according to St. Paul; but scientists are certainly within their rights to ask what it means to be divine anyway. Thus, here is where it gets tricky. To understand the mind of God is to understand his intelligence. But to understand his intelligence is to understand his divinity. As a result, a scientist will stop and probably scoff because he or she can only measure or observe that which is physical; divinity by definition is not physical.

For starters, we get confused when the Bible itself points out that we humans are created in God's image. But I challenge anyone to tell me what God's image is. Image usually refers to something defined by the dimensionality of space and time; but we will see here that God must be beyond space and time. So how can he have a physical image?

God's image must refer to something much more profound than that which makes our bodies a physical entity, hence the argument that God's mind does not have a parallel with our minds. Perhaps St. Paul had a handle on this concept when he wrote to the Corinthians that "no one comprehends the thoughts of God but the Spirit of God"; likewise, no one comprehends our own human thoughts than the spirit within us, he goes on to say. Thus, in this way of thinking, the image of God is replicated in us; it extends well beyond our physique, into another realm, perhaps another set of dimensions that are even wrapped around our own physical dimensions, and maybe more akin to spirit or soul or common consciousness. While leaving this distinction for the theologians to contemplate, let's stick with consciousness for now.

In cosmology, it is perhaps the "in thing" these days to talk about the driving intelligent forces of our universe, or parallel universes that could exist, or even different levels of consciousness that could exist inside them, and that consciousness can be similarly dimensioned with the rest of the hyperspace that makes up our own universe, which includes the normal set of dimensions of space and time. We'll see later how these dimensions become involved in a possible answer to the question of how to regain the symmetry of the universe, and its theological meaning for understanding God and our relationship to him. Some theories point to 10 dimensions, others as many as 26! Well this is all very interesting, but you need to understand that physicists will always derive things from a mathematical point of view. So when he or she talks this way in a scientific discussion, what he or she really is saying is that there is a singularity, possibility, or even finite probability that an event can occur within the theoretical framework of some set of equations that try to explain our universe. Problem is, no one seems yet to agree on grand unification theories (or GUTS as physicists call them) or a Theory of Everything (essentially a kind of GUTS with gravity thrown in) because they haven't been renormalized yet to anyone's satisfaction. And the reason for that is that not enough physical data exists to understand the relationship of the fundamental forces of nature (gravitational, strong, electromagnetic, and weak) into one single unique beautiful equation to describe our universe; most of these equations basically "blow up" at some point because they have terms that go off to infinity that are not easily canceled. You have to realize that the first symmetry breaking of the universe occurred when the fundamental separation of gravity from the other still unified forces of nature occurred sometime around the ripe old age of 10^{-43} second after the absolute moment of creation (Big Bang being $t = 0$ if you will) of our universe. And that's a mighty short amount of time to be getting data from, much less to be there with a stopwatch to collect it in the first place, especially considering the universe has aged about 13.7 billion years since then, plus or minus an eon or two. We'll see later why this brief moment after creation became so important to physicists and why it might just also be of theological interest to the rest of us; but suffice it to say right now that this number didn't just come from nowhere.

The point though is that there seems to be a common consciousness that exists in our universe; and it prevails over the universe, encompassing our own as if it seemingly belonged to a collective. It's a "did the tree make a noise when it fell in the forest with no one around" kind of argument, and thus doesn't yet really satisfy anyone's thirst for knowledge of the origin of our cosmos, how we got here, who or what is God, and what is our interface to him. But we haven't gotten that far along yet; so be patient. For the moment we continue with our consciousness and its relationship to God's intelligence, the mind of God.

One way to understand it is with some examples. When Africa cries out from its hunger or those dying of AIDS, famous rock bands the world over combine their talents to perform and raise money for the cause; and it motivates the rest of us into action. When natural disasters hit communities, such as 1906 San Francisco or 2005 New Orleans, the mind of God leads all of us to unite, bring aid, and help them rebuild. When we see America use its industrial might to bring compassion and help to the world, the world community tends to rally behind its Big Brother neighbor. On the other hand, when America shows greed and naive impulse to push smaller neighbors around and then abuse prisoners at Guantanamo and Abu Ghraib, the world reacts with blistering protest and resentment against its leaders. And when Germany was annexing Europe during World War II and gassing people at Auschwitz, the world reacted with utter revulsion and resolve to bring Adolf Hitler down. And one can see why this all happens. Just ask Muslims worldwide how they feel when they hear about someone flushing the Koran down a toilet. The world reacts with a general overall collective common consciousness of goodwill for the good, disdain for the bad, and repulsiveness for the ugly. The common consciousness that we all share brings us into union with the driving force behind creation and causes the collective to react the way we do. Hence it is a postulate here that there is indeed an organized intelligence and purpose behind the driving force of our universe which is not human in any way, but which allows human access via the collective shared consciousness of the world community, and indeed all intelligent life; and it becomes effectively a window that peers into the mind of God. Instead of waving this shared consciousness off as just something we cannot explain, and thereby simply assigning God to it as the famous late physicists Richard Feynman has suggested we humans are wanton to do, we actually make an attempt at a theory of explanation, after we borrow some ideas from string theory later on.

From a Christian perspective, it could be understood what Jesus meant when he said that when two or three are gathered together in my name, I am there also. Through the shared fellowship of the two or three gathered together, each can see directly through an effective window to the mind of God, by sharing in the group's common consciousness in community, fellowship, worship, and celebration. From an Islamic perspective, there is no fellowship between humanity and God himself, except to have sole obedience to him whom nobody can really comprehend; however, Muslims believe in maintaining a social community to enhance the submission and unquestioning loyalty of the community to Allah. So perhaps the importance of this fellowship in all our faiths is revealed to us by God through our activeness in our various places of worship, in mosques, synagogues, and churches. It may also provide a reason for the phenomenon we all experience at rare moments in our lives when we are thinking of long lost friends at just the same moment they are thinking of us.

Through a shared consciousness in our fellowship with each other, we help weed out the individual deviations, biases, and greed of our own personal desires and wishes. It allows us therefore to be able to realize what God is "thinking" as it were. When we act out of selfish desire, greed, vengeance, or arrogance, the rest of our community responds with indignation and distaste. One can even project this idea to the family of nation states reacting to the rogue actions of individual nations. In this case, the aggregate community of nations can see through a single nation's pride or prejudice when it denies its own access to the mind of God, which can only be seen then through that of the larger world group of nations. And usually, the end result is an act of God that turns the tables on the violators with a "taste of their own medicine", a kind of "eye for an eye" if you will.

One can see this point with two parallel examples. When World War II started in 1939, German troops dressed as Poles in order to stage a takeover of the Gleiwitz radio tower, just inside Germany near the Polish border. They did this because the Nazi government needed a reason to satisfy the people of its desire to invade Poland. It then announced over the radio station (I'm not kidding) that they had done this. Well the government was only too happy to tell the German people what was happening. It worked. Germans endorsed a war with Poland with blind nationalistic fervor. Those who willingly stood up to oppose Hitler in Germany, such as Dietrich Bonhoeffer (1906-1945), a prominent Lutheran theologian, moral leader, writer, and pastor, were summarily sent to the concentration camps, and along with the Jews were martyred. Yet the world could see all along what was happening; and it responded with indignation and despair. In the end, it would be the German nation that was conquered and the leaders of Germany who got a taste of their own medicine with the loss of their own lives; but the mind of God in the worldwide community had prevailed. Today, it is quite refreshing, for instance, to see a German pope coming back to Germany and lending support and strength to the Jewish community!

If any of this tail of political deceit and aggression seems to have a less extreme yet more recent parallel ring of familiarity about it, fast forward to 2003 and America's invasion of Iraq. According to a document, now known as the Downing Street Memo, that was drafted by the MI6 intelligence service in Britain and made public during Prime Minister Tony Blair's third reelection campaign, the U.S. White House decision makers needed intelligence to fit their policy already in place to invade Iraq; in the words of the document, "intelligence and facts were being fixed around the policy". As if torn from the German playbook, they propagated a story of Iraqi Weapons of Mass Destruction that the world was soon to see as Weapons of Mass Deception, even to the point of seemingly deceiving their own messenger to the United Nations - the honorable and accomplished U.S. Secretary of State Colin Powell. From a "Chinese menu" of deceptive phrases given to him, he was to tell the U.N. that Iraq had "active chemical munitions bunkers". Again it worked, as a great majority of Americans initially responded in favor of the war; but the world didn't buy it, as it protested in despair. Those in America who were opposed, such as former U.S. Ambassador Joseph Wilson, who had worked for the National Security Council and had done investigative work for the Central Intelligence Agency, were scoffed at while their credibility was put into question. The world knew what was wrong before those in America who were caught up in a blind nationalistic fervor that something was wrong. Maybe the person who coined the phrase that pride is the root of all evil was really onto something after all, evil being that which is opposed to the mind of God. As did the German people before them, Americans repented and despaired along with their world neighbors at the knowledge of what their government had befallen them. Eye for an eye, it became the leaders in America whose reputations would be brought into question by the world; but the pervasive mind of God had prevailed. Today these leaders still worry about new insurgencies and civil war in Iraq, while totally oblivious to the fact that they themselves were the cause of it; but once again, it is refreshing to see Americans finally come back to their cherished freedom of expression to protest and demonstrate against wars built on false pretenses.

Now when a country uses religion to spread terrorism or war around the world, it has to know that it works in opposition to the mind of God. And when another country tries to stop it, it has to know that it loses credibility when it lets its own clerics pray for the death and assassination of others. Similarly, when organized religion uses its political strength to create national violence, it works in opposition to God. And when organized religion tries to stop it, it too must know that it loses credibility if it rallies around leaders who promote war. God is serious when he says "Vengeance is mine, I will repay"; and when you decide you want to play God using your own rules, you run the risk of the wrath of God. After all, God could easily say "Who put you in charge?" Unless you have the mind of God behind you, as seen through the common consciousness of us all, you may find out soon enough what the expression "an eye for an eye" is all about.

On an individual level, what happens to us when facts are purposely mislaid or restated simply to fit our wishes; but when they are exposed, we repent. Schoolboys will see the lie, scholars will see the cheating, researchers will see the bias, and great leaders will see the crimes against humanity where war is involved. It is that consciousness that we all share with each other that is easy to see when we are in communion with each other, but hard to notice when we are alone or with those who blind us with their ideological motive; and it allows us to peer directly into that which serves the mind of God, a mind that preaches truth, peace, and love for the world. That which is opposed is rooted in evil.

So it can be recognized that when humans commit crimes, pass judgment, or enter into wars against other humans, you can find whether God supports you or not just by reading the shared consciousness of the world community. Whether it be the ongoing Christian crusades against the Middle East (Middle Ages to the present), Islamic Jihad and terrorism against the West, Jewish occupation of the West Bank, or Palestinian and Iraqi suicide bombers who have a personal, political, or religious vendetta, the world has shown through its conscious protestations that God has declared it to be wrong! It's not what he wants. And those who take part in these actions, regardless of religion, simply try to justify their actions for their own ends; and the world responds and thus shows that God is aware and not pleased. He responds with peace, and not war; if he did otherwise, Jesus for instance would have had an army to support him. But he does use a more insidious weapon; and that is our consciousness. Again, we'll speculate later just how he may be creating the interface to use this "weapon", where we borrow from concepts used in string theory.

Notice that all intelligent life is included in this shared consciousness, because as hard as it is for us arrogant humans to admit it, animals like our dogs and cats have consciousness too, albeit at a reduced level; thus their road to heaven may well yet be paved! Good thing too: the thought of a heaven with only humans is too horrific to bear! But it makes sense that animals share in the mind of God by the behavior they exhibit. I think back to the last dog I had, and how loyal, faithful, loving, and kind she was; there is no doubt in my mind that she also shared in the mind of God. And it is really no surprise to me that in the world's most spoken language, "dog" spelled backwards is...well you get the point! But animals also do ugly things such as fighting; but you then realize they only do this to maintain their survival instinct and obtain food (as humans will do as well, if tragic events in New Orleans are any indication!). And this may be pushing it, but you also get the feeling sometimes that even animals know when it's wrong. When we speculate later just how God may be entering into the conscious of all life forms, through physical consciousness, we'll also be able to see why it is that humans have an easier time recognizing it. I guess it makes sense, however, that it is part of God's plan to speak mostly with humans; after all, the after dinner conversation would certainly be more meaningful and interesting than it would be with a cat! Yet that isn't to say that God cannot or does not want to speak to your pet. Remember that next time your leg is being seized because your pet wants attention. God needs attention too; and like your pet, if you don't give it, there's likely to be trouble! But this comparison only goes so far, because the question is likely to be asked, whose master is who?

And that brings up a side point, the way in which we treat our furry friends. When an animal is abused, it suffers like any of us would. When it is not cared for, it hurts too. There is a reason for this, beyond the obvious. When our conscious speaks to us or their conscious feels pain, just guess for a moment who is being the mediator! When our common consciousness peers into the mind of God, you can bet that God uses our conscious to peer into our minds, and those of his animals of creation. If you decide to treat an animal harshly, and not as you would do for yourself or those you love, be careful just who it is you are really hurting. For Muslims, the command of the Koran comes to mind that we are to be kind to animals. For Christians, the words of Jesus come to mind: as you treat the least of these, you also treat me. Theologians will be quick to point out that Jesus was referring only to the poor among us; but it would behoove us to consider that God gave animals the gift of consciousness and feeling too, and probably with good reason.

Another facet of this shared consciousness could be a speculative ability to communicate with those that have passed on beyond their life in this physical universe. Now this idea really gets out there with all kinds of speculative ideas about magic and karma, because there would never be any physical evidence to explain the phenomenon. A distinguished scientist wouldn't touch these concepts with a 32 foot pole for fear of being called a medium; but that won't stop us from trying it. Now, mind you, there is no suggestion here that you should run down to your neighborhood psychic to find out about your future or to tell your long lost loved one about how much you miss them. What is suggested however is that even though we know that our physical bodies die because they are dependent on a universe of space and time that is in itself finite, we may well exist in one or more other dimensions (that even physicists admit they are trying to model), such as a dimension that could record our consciousness, for instance; and it would not necessarily become lost as our physical bodies will one day be lost. And if our consciousness can exist in another dimension that doesn't depend on our physical brains for existence, then our dead loved ones are not necessarily entirely "dead" after all. And if they still exist in another set of dimensions, then there is no reason we cannot be there also. Therefore, it would be possible for a connection to be made. As we remain present in this universe in the physical dimensions of space and time, it may become too difficult for us to peer into another set of dimensions; but the possibility is not written into the code of physics that some dimensionality of our existence could not do so. What is interesting here is the speculation that our physical bodies and brains could have some interface to other dimensions which would be independent of our biological necessities of physical life. But that does not mean you should run down to your friendly neighborhood psychic, become some frauds do exist and some people are just full of fairytales; on the other hand, some people may truly have a gift of consciousness that allows them to see more than the rest of us. Most of us probably have friends that have shared some supernatural experience with us that they can't explain. Without necessarily dismissing their experience to some psychotic state of mind, one could reason that conscious may have played a role in these experiences because it has some possible hook or interface to other dimensions beyond space or time.

Of course, you can possible see now where this is all going. If we talk about things other than the physical, then we must be talking about things that are spiritual. Well this is not necessarily true. Remember that we were leaving it to the theologians, for now, to explain the difference between the soul, spirit, and consciousness; the distinction however will come up again when we discuss string theory and God's interface to our conscious. Then, we'll see that consciousness, by sheer coincidence, gets modeled geometrically by psychologists today in a very similar way that string theorists are modeling other hyper dimensions (other than space and time) of our universe; and these dimensions give rise to the heretofore lost symmetry of the universe, a place where a perfect God of symmetry could reside, simply because he would have been around before the asymmetry of the universe ever occurred in the first place. Thus, our interest lies in trying to piece together these theological, physical, and mathematical concepts which can give us clues about the nature of God from a place we all can appreciate and understand, the space and time of our own universe.

Then there must be a driver behind all of these dimensions, and the ability for our universe to function; it serves, and can be seen as coming from, the mind of God. We have access to it through the shared consciousness of the world community which proves its existence, because the universe is coming from a clear starting point and heading in a certain direction that sets the collective community of beings (human or otherwise) in motion; and it is not simply just going through a perfectly random series of events in the limited dimensions of space or time, but is in fact an organized existence that we as a community have the pleasure of experiencing and sharing. And it is through sharing as a community that we can even gather some information on the personality of God. Most religions believe he has anger, pity, love, and kindness, detectable in the way he decides to communicate with us. All these emotions occur biologically within our own consciousness also; and that probably is not just a coincidence. But we can also talk about one of the maybe least understood aspects of God, the humor of God; and yes I do mean humor. We can see this aspect in the general joy we share with each other in fellowship, whatever our religion. Christians for instance can point to the personality of Jesus, and the many good times he had with his disciples. In fact, in the most profound moment of human history, Christians tell the story of a resurrected Jesus facing a crying Mary in front of the empty tomb, as she asked where he had laid the body, thinking Jesus to be the gardener. What was Jesus' response, but none other than to say "Whom are you seeking?" (as if!). Now, that was a good one; but clearly he must have been anticipating and readying himself for the joy of the moment.

Up to this point though, we haven't made the connection between the consciousness we all have, that gives us the shared ability to peer into the mind of God, and exactly just whose mind it is anyway into whom we are peering. It's like the unknown person who comes out of no where to win the gold medal; you just want to exclaim: who *is* this guy anyway! If you accept that he exists, then how did he get here, and where exactly is he taking us? For that, we have to pick up the clues left behind by our universe itself, and try to understand how the universe came into being in the first place. From there we can begin to understand what God is all about, appreciate his majesty of creation, and what it is that separates us from him anyway, beyond the usual theological answer that has something to do with humanity's woeful state, due to its own weaknesses that separate us from the perfect and almighty God. The universe itself has left behind these clues from creation up to the present day. Cosmology and physics for instance have really only just begun to probe these clues during the last 100 years or so; and they may go a long way in helping us to understand some of these questions, including even a possible theoretical approach to spirituality, such as Jesus' divine reappearance to Mary.

Perhaps this is as good a time as any to bring up a rather curious but not forgotten second century A.D. Egyptian religious and mythical philosopher named Valentinus, who founded the Roman and Alexandrian religious schools of Gnosticism which taught that God contains all things within himself, including the cosmos which was a subset of himself, able to surround all things while nothing is able to surround him, inconceivable, uncontained, and superior to all thought. Legend has it that Valentinus studied philosophy in Alexandria under one named Theodas, a purported pupil of St. Paul of the biblical New Testament; and he was baptized as a Christian. He went to Rome around 140 A.D. during the pontificate of Hyginus with the idea of becoming the next bishop of Rome – nice idea. But when he was rejected to be the ninth successor to St. Peter in 142 A.D., apparently by a very close vote to Pius who became pontiff, he decided to leave the church all together to establish his own brand of religious mythology, but curiously not before Anicetus took over from Pius in 154 A.D. By 160 A.D. though, he had returned to Cyprus and possibly Alexandria to continue his own teaching, and died shortly thereafter.

What is interesting about Valentinian mythology is that he created a kind of cosmogony which supposed that an original nothingness, where space and time is nonexistent, corresponded to an original infinite and indescribable Godhead. The infinite symmetric nothingness is good and any breaking of this symmetry (for instance the modern physics predicament of the rise of matter over antimatter, discussed in The CPT Conundrum) represent imperfection and evil because of its finiteness and asymmetry. The human fall from grace is characteristic of our place in the imperfect universe. So far so good, as far as Christianity, Judaism, Islam, and science are all concerned. Unfortunately, Valentinus went on with his ideas. At the center of his Gnostic thought was that not only are the universe and human existence wrong and imperfect, but that the Creator himself is to blame for it all. His version of creation begins with a primal being called Bythos who remains quiet for a long time, and then suddenly calls out to 30 beings called "aeons" (also known as the "pleroma"). And it gets better; apparently one of the lowest aeons, Wisdom or Sophia, becomes pregnant and gives birth to a demiurge, Jehovah, who in turn creates the physical world. It's hard to believe that this guy (Valentinus) almost became one of the first Catholic popes; perhaps his bitterness in not achieving this office played into his thinking. In any case, you have to give him credit for his imagination! Valentinus was also a prominent mentor to one Ptolemy, who had the original concept that the earth was the geocentric center of the universe, which would be taken up and held by the church as a theological truth for centuries. As we will see later, this wouldn't be the only idea from Greek mythology that the church would wind up borrowing as a theological truth and then make out to be a fundamental principle doctrine of the church which would stick around for centuries, except that in this case, it has bound and split the church even up to the present day; and we'll find the mastermind of it may have been none other than Valentinus himself!

In 1945, as it turns out, buried in the sands of northern Egypt, a series of Gnostic "gospels" were found, and at least one of them, The Gospel of Truth, is attributed to Valentinus. In it, he makes Jesus out to be a metaphysical philosopher in his conversations with his disciples; but the funny thing is that he only interprets passages from the already accepted Christian gospels (Matthew, Mark, Luke, John) so that there is no new factual information put forward about Jesus' life. Since the second centurian Valentinus could not have been a contemporary of Jesus, the conclusions he reaches seem only to add merit to his own fantasies, and not add any real insight into the history of Christianity.

However, one could reasonably theorize from Valentinus' view of the world that the whole reason the universe is imperfect, and thus came into existence in the first place, was not humanity's fault. If it was humanity's fault, why was the universe created with so many imperfections at the start when humanity wasn't even around? After all, we know the universe was created from a basis of asymmetry of matter over antimatter; and its ultimate demise is connected to the way it was created in the first place. Could it be because the universe was meant for the expulsion from heaven of those that did not obey God? Could the asymmetric and imperfect universe be the result of the first space battle of all time, when only the heavens existed and the archangels Michael and Lucifer, with their legions of angels, fought for control? Lucifer was ultimately cast out of heaven for trying to take over the reigns from God, as described in the biblical book Revelation, and Michael was the victorious general in charge, as it were. (After all, my middle name isn't Michael for nothing!) As for God living in a place of perfect harmony and symmetry with everything around him, he would have had to find some place distinct, separate, and imperfect from the heavens to cast Lucifer into, if you think about it. So then, we're told in the Bible that Lucifer, now called the devil, was cast upon the earth and hence left for humanity to have to put up with. As one outcome of this story, Michael is revered by the Catholic church even today, where the faithful are supposed to ask him to save them from the battle of evil and sin brought on by the devil. And Lucifer, the once beautiful and majestic archangel who had a strong voice and a charismatic following, is now the most jealous, rejected, and despised Being of all time.

Well, the nice thing about this kind of reasoning is that we humans could be off the proverbial hook. Namely, we cannot be held responsible for the imperfect universe that ultimately leads to our own downfall. Unfortunately, even if this is the case, the church throughout the ages has gained a lot of mileage by pronouncing judgment on humanity and making man the cause for his own peril, a kind of guilt trip really. The speculation offered here as a possible reason for our own creation would suggest that this really should not be the case, namely that God had some other purpose in mind when he decided to code the universe with the feature of finitude, if you can call it a feature anyway. (In software engineering, one man's bug is often another man's feature, and vice versa; it depends only on motive really!) Thus, whether you buy into Valentinus' idea about the "fault" of God for our imperfect creation, or you take the idea behind the great war in heaven as giving reason for the birth of our universe, you would logically reach the conclusion that man isn't responsible for his own mortality. And chances are, you would still be in disagreement today with how the church, as a whole, sees our sorry state.

Frankly, it would seem that Valentinus has the weight of scientific evidence, at least on this point, on his side. Because we know that an initial symmetry breaking of forces, which allowed the majority of particles over antiparticles to form our physical universe (and as we shall see causes its eventual demise), led to our finitude, how could humanity be responsible for its eventual fate when it wasn't even around when all the symmetry breaking occurred in the first place? It has been suggested by some metaphysicists that if we are ever to understand the details of the process that gave rise to our physical universe and its ultimate demise, we may be able to use images from mythology, such as those from Valentinus, to help us along. And knowing this process, what would that say about the restoration of symmetry that would be required if we ever hope to regain our own reconciliation to God through salvation, spirituality, or our own consciousness, a consciousness that provides the bridge already to the mind of God.

There are many other arguments of course that one could make for the reasons why we have our existence as part of the creation of the universe; and physicists certainly have their share of theories. However, the elephant in the room remains; who is this creator and how does he bring about our, and his, creation in the first place. We now have some idea into the where, what, and when, as far as our own universe is concerned; but theologians are tormented by the who and physicists are tormented by the how, because all the known scientific data in the world will still not answer the fundamental question about what came before space and time to begin with, and what brought about that ultimate singularity, called euphemistically these days as the Big Bang, that gave rise to our universe in the first place, and the asymmetry and imperfections that came with it, as it turns out, that were necessary for our own existence.

Well, we can start easily and naively enough about the who by checking with the Bible. The Jewish God of the Old Testament of the Bible sometimes comes across as a warrior-like God, for instance. And Christians who adhere to the New Testament struggle with this interpretation, especially when they consider their perception of Jesus, the Son of God, considering what he said and did; for them God seems to have a split personality. But I'm here to tell you that the God of Judaism is the God of Christianity is the God of Islam, if the biblical genealogy of Abraham is to give any indication; and he is the same now as he was at the creation of the universe, if the laws of physics are also to give any indication. A friend of mine likes to say that the Bible is a living progression of our continuing understanding of God. That's true; but the problem here really lies in just who is telling the story in the first place. As you might find in any culture, the God of the Jews in the Old Testament needed heroes and leaders. Of course, along with the heroes and leaders come the dummkopfs and losers. And a large part of the Old Testament describes just these kinds of clashes between the two. The writers primarily belonged to the various tribes of Israel. Funny thing was that not only did God seem to approve the catfights they had with neighbors, you usually can guess by reading the Old Testament with whom God is going to side. In one obvious example, it is written in Malachi that God "loved" Jacob, but that he "hated" Esau; yet both were early fathers of neighboring nations. Well, you can guess from which side of the fence the writer of Malachi hails! However, if you look more carefully, these writers quite often described the personality of God with a much higher quality, grace, and dignity. He could be angry, he could be sad, he could be happy, he could be humorous, but he was always charitable, caring, and loving, characteristics that Muslims find in the Koran with Mohammad, the prophet of God, and Christians find in the Bible with Jesus, the Son of God. Where God does seem different in each religion is mostly tied to humanity's relationship and interaction with him. God is the paternal parent of Judaism, the revered and feared of Islam, and probably the most approachable of Christianity. And here is where a few more traits probably show themselves. Yes, he's a genius and very wise, but he also has a jovial and gamely side to him also. We see this in the way he set forth the rules of creation, and in the way he decided to make everything work. Sometimes, the rules are pretty; but sometimes they are down right nasty. And physicists are still trying to understand those rules, likely with all the blessing and bemusement of God looking on, although they probably don't prefer to see it that way.

In order to gain a more fundamental understanding about the who of creation however, we have to switch over to the question of the how of creation. And that's where the physics part of this discussion comes into play. And if physics was necessary for the creation of the universe, then there must have been a physicist (and maybe software engineer, as an aid!) who coded the laws of logic of the universe. Thus, here is where God the physicist, and the how part of the story, begins – with creation, the process and tools that brought it about from what we do know and what we don't, the way we came to know it or not know it, the universal asymmetry that resulted, what brought about our finitude, where we are headed, what we can do about it versus what we can't, the dimensions of the universe where universal symmetry and thus infinitude may still be preserved, and how we could ever get back to that point. We shall show how the loss of symmetry in the universe has brought dire theological implications for humanity; and we will see what it would mean to get it back. It may be perhaps through this effort that the grace of God would allow us to regain favor with God. But to understand it, we have to explore what it is that God created in the first place, how it came about, and what exactly is happening to this universe we are forced to inhabit and share. An understanding of these topics would then give us a clue about the real who and what that God is, and what is in it for us as part of our own relationship to him.

V. The Relative Creation

When Isaac Newton was developing his universal laws of gravitation, he would sit up at night gazing at the moon. I'm guessing he had more than his share of sleepless nights. But I guess he had to make the change in his sleeping habits; they say he almost went blind by glaring at the sun too much during the day! But what really puzzled him was why an apple could fall to the ground but that huge rock in the sky didn't come careening down into the earth, both of them actually. One wonders how he would have responded if somebody had told him that the moon had actually been carved out of the earth when it was about 50 million years of age or so. His eventual equation that linked the force of two orbiting bodies about each other, as proportional to their masses multiplied together divided by the square of their distance apart (the proportionality constant being Newton's universal gravitational constant denoted by G), was extracted from his measurements of the planets' motion in the night sky. What he never accounted for however was *why* they interacted in this way. Physicists thought for centuries that space was simply a medium on which forces depended to find their relative strength at any observed point. It certainly seemed true of the electromagnetic force when measured between charged particles acting on each other; again the force was in proportion to the inverse square of the distances they were apart. What no one had even bothered to consider was that gravity seemed to be a different animal all together. It wasn't just that gravity could be measured by its dependence on the distance of space between two objects of mass; it was that the space itself was dependent on that which caused gravity! And then, who could have imagined time would also be dependent on gravity. Mind you, we're not talking about the ability to do time travel here, although it is probably where those **Star Trek** writers got their big ideas from; we are talking about the ability for gravity to seemingly slow time down. So what gives rise to gravity? For this, we need the help of one of the greatest physicists of all time.

In 1905, Albert Einstein, while working for the Swiss Patent Office in Zurich (he couldn't find a job as a physics instructor!) worked out a series of equations that link space with time, where time can mathematically be considered a 4th dimension of our 3 dimensional space, that all physical things can not travel faster than the actual speed of light (186,000 miles per second), and that funny things start to happen as mass or energy starts traveling at speeds close to light speed. Light speed did not depend on time or place; and it was a constant quantity regardless of the inertial reference frame of the observer. He called it the Special Theory of Relativity; he did it aside from his published work on the photoelectric effect that he was doing at the same time, for which, it would just so happen, that he would end up receiving the Nobel Prize in physics in 1921.

In relativity theory, no mass can actually reach the speed of light without first turning into a massless state, or, equally unpleasant, needing infinite energy to do so. In fact a particle mass m has a relationship with its energy E (remember $E = mc^2$, c being the speed of light) when it is treated like a wave, as we shall see that quantum mechanics suggests it can be. But maybe the most interesting thing about this theory was what it said about the relationship of reference frames moving with respect to one another. For instance, a spaceship moving at near the speed of light would have an onboard clock that goes much slower than the clock of the observer on the ground; and distances would appear greatly reduced on the spaceship compared to the observer on the ground. The theory is a mathematical formulism based on the famous Michelson and Morley experiment done in the late 19th century which showed that the speed of light was constant regardless of its origin reference frame (in the case of their experiment, light was measured parallel to and then perpendicular to the motion of the earth around the sun), and that the same laws of physics apply in any reference frame of measurement. Thus the theory of relativity requires that the speed of light is a constant in any inertial reference frame and with time; we'll see later how important the word inertial becomes when applying relativity to the expansion of the cosmos. For now, understand that the phrase inertial frame refers to any frame not acted upon by external forces.

Now I should mention as an aside that a few present day theorists have actually taken issue with this assumption about light and the Michelson and Morley experiment itself. They say the paradox of being able to make the same measurement of light speed for someone moving toward a light beam versus someone moving away from it could in fact be subject to classical mechanics, namely that the speed of the beam would have to take into account the speed of its reference frame of measurement. The reason is, they say, because Michelson and Morley failed to account for the travel of light going back and forth in their laboratory of measurement, because the effects of the earth's reference frame would be cancelled out due to the number of equal times that light was going in a plus direction as it was going in a minus direction along the axis of measurement. So far, there haven't been many takers for this argument, but stay tuned on that one.

By 1912, Einstein was finally recognized for his brilliance and became Professor of Theoretical Physics in Zurich. In 1914, he moved back to the country of his birth and gained his German citizenship (as he had spent most of his young life in Switzerland), and became Director of the Kaiser Wilhelm Physical Institute and Professor at the University of Berlin. Having never let go of his Special Theory of Relativity, he formalized it in 1916 to include gravity; he called it the General Theory of Relativity, which would outdo his first masterpiece on Special Relativity, and still be in use today. It is a mathematical formalism that essentially allows one to calculate different fabrics of universal space and time and the relative relationships of mass bodies with respect to each other in this space and time. It postulates that light speed is constant under any reference frame at rest with respect to gravity. The theory has become the greatest contribution to our understanding of gravity since that apple hit Sir Isaac Newton's head some 250 years before. (Well, it didn't actually hit his head, but the story certainly sounds good enough.) Without it today, we could not do GPS or time precision satellite technology. And hence you can see why a GPS-less SCUD missile never really knows where it is going, whereas a Tomohawk missile will hit its mark, never mind those few times Saudis complained that Tomohawks were seen landing on their desert during the wars in Iraq; however, when there is an error today in GPS timing, it usually has to do with the on-board real-time clock or operating system more than anything to do with the physics.

Under Newton, there was a universal proportional gravitational constant that allowed one to calculate the force on two planets by each other that was equal to this constant times each planet's mass multiplied together divided by the square of the distance between the two planets; in this scheme, the universe was homogeneous under the force of each planet's gravitational attraction. Only problem was that Newton failed to show how each planet knew the other was there. But Einstein pointed out that the universe is in fact not actually homogeneous, but is warped in space and time by each massive star giving rise to the effect we call gravity. Gravitational fields effectively produce ripple effects in the space and time continuum. Time is still a 4th mathematical dimension of the 3 dimensional space we live in, as it is described in the Special Theory of Relativity; but gravity is not really a force at all. And here's where things really start to get interesting. Gravity is the effect of mass warping the space and time around it (like a bowling ball on a waterbed) to create this mysterious attraction of a smaller mass to it, giving the effect of a gravity well, so that even light itself would be affected. Now you know why your feet are stuck to the ground on the earth, and not wondering off into outer space somewhere; and good thing too…your lungs might not take well to the vacuum of outer space.

Now there were important consequences that the General Theory of Relativity predicted about light itself. For instance, light propagation and speed could be affected by any new gravitational field it encounters; and it would lose energy and thereby have its observed wavelength lengthened (redshifted). Planets would not just simply orbit each other due to their gravitational effects on each other, as explained in Newton's theory, but they would also precess around this orbit (shift the orbit slightly) each time around. The precession of mercury's orbit was a prime motivating factor and boundary condition in Einstein's development of his General Theory. Physicists for a long time were puzzled by mercury's motion such that they devised a theorized planet named vulcan (I'm not kidding) that must have been orbiting mercury in some way to shift its gravitational motion. Only problem was no vulcan was ever found (too bad for Spock). So after Einstein formulated his theory, he was careful to do the calculation of mercury's precession and found to his amazement that it matched the observed value almost exactly. His colleague Max Planck (for whom the Planck constant and Planck length are named) once told him (paraphrasing): you know, if this theory's correct, they'll call you "the new Copernicus". Copernicus indeed! Yet Einstein never had to write his own treatise to honor of some pope! But that didn't mean the pacifist Einstein didn't have his own fair share of politics to deal with. We know of course of his disgust for Hitler, who rose to power at the time when Einstein was still in Berlin; and it seems Einstein had a friend and physicist named one Friedrich Adler, an assassin who had just decided one fine October day in 1916 to shoot the Prime Minister of the Austro-Hungarian Empire. Einstein would later rally to Adler's defense as a character witness.

Enter Arthur Eddington (1882-1944), an English astronomer at Cambridge who had traveled to the west coast of Africa to measure the traversal of light from the Hyades constellation with photos he would take during a daytime total solar eclipse, that was happening there on May 29, 1919. He had become interested in General Relativity because it had explained the precession of mercury's orbit. So he arrived in Africa by boat in order to observe the theory's predicted bending of light by stars around the sun, which could only be seen at that time by an eclipse. Unfortunately, on the day of the eclipse, it had decided to rain! His team waited anxiously for a break in the clouds to occur; it finally did, and just one-half hour before the eclipse began! He got his photos and stayed there to compare them to nighttime images of the same constellation. He discovered that the daytime photos revealed a deflection of 1.6 seconds of arc of light from the stars whose light path to earth was closest to the sun versus those that were taken at night, thereby making one confirmation of Einstein's General Theory of Relativity that heavy gravitational massive objects warp the traversal of light. That result would make Einstein instantly famous in a world recently exhausted from world war. The idea that light, being electromagnetic radiation without mass, could be affected by the gravity of planets was amazing. And for Einstein then, what did this say about the interaction of gravity and electromagnetic radiation; and how could this be used to form a universal theory, or explanation of everything, which could be tied together with all forces of nature that he had desperately wished to find? Unfortunately for Einstein, he would never get that far; but by then, he had predicted something that would cast him onto the world stage at the same level next to Newton, much less Copernicus.

The story for Einstein continued to get better as Edwin Hubble (1889-1953) discovered in 1924 that more distant galaxies from earth had a systematic redder complexion than nearby galaxies. This is known as the Doppler effect in physics, which suggests that light downshifts toward a redder (lower) frequency when items are moving away; the faster they move away the redder the shift. Thus, another enormous triumph of the General Theory of Relativity was established which had as one of its predictions that the universe should be collapsing or expanding, and more distant objects would recede more rapidly, like dotted objects on an inflating balloon, if in fact the universe were expanding. Unfortunately, Einstein didn't want to believe what his own equations were telling him because it seemed to him that the universe should be static; after all, that was common sense. Otherwise, why didn't the universe collapse long ago under its own weight of gravity. We'll see below how this led to the biggest "blunder" of Einstein's career, and how and why cosmologists today have resurrected this "blunder".

So far though, you would be right to guess that the General Theory of Relativity has held up well to all scientific investigations and data taken to date in our cosmos, except for those at the tiniest quantum mechanical level or at the farthest reaches of the visible cosmos, but more about that later. What had really frustrated Einstein to the end, and all physicists since, was the desire to put all the fundamental forces of nature together into a sweet unification scheme alongside his General Theory. Einstein couldn't do it because not enough information was known yet during his time. It's been said that if Einstein were alive today with the advances in our knowledge of the known forces of nature (strong, electromagnetic, weak, and gravity), he would have connected gravity with Grand Unification Theories (or GUTS as physicists like to call them) of the universe, and maybe found the final Theory of Everything. While Einstein never finally "got there" with his final unification scheme, he did unleash an incredibly successful and enormously powerful theory for 20th century science to ponder that has created repercussions in everything we now understand about our universe. However, after Hitler took over Germany in 1933, Einstein renounced his German citizenship and came to America to become Professor of Theoretical Physics at Princeton, where he would retire in 1945, just as Hitler was being defeated; but for a brief moment in 1933, Berlin must have seemed the center of the universe, housing leaders that threatened the world and science, as we knew them, all at the same time.

The one item that perhaps doesn't get proper treatment however, except to cosmologists, is the notion that not only space *but* time is also warped by physical mass. Wow, Aristotle would have had a field day with that. As it is, Aristotle thought the effect of gravity on an object was in direct proportion to its mass, and that things would fall infinitely fast in a vacuum; but of course he also thought women had fewer teeth than men. So no help there from one of the leading philosophers of all time, but in his better moments, Aristotle had thought movement is eternal and change of movement implies eternity because the fact that change can occur means there was no beginning or ending in time. Otherwise, there had to be a first or last change, which is impossible. He went on to say that the source of all movement must therefore be God who is the prime mover of the universe and whose only activity is thought. Aristotle may have finally been on to something here. Stephen Hawking might have agreed with him when he argues that if space and time have no boundary, then there just might be no beginning, or moment of creation. However, time would have had to be something that was created at the moment of creation, along with space; so to talk about time before Creation has no meaning, since it simply did not exist. For an observer (e.g. God) outside the box of creation, both space and time would have no meaning or dependence, because his or her existence in an outer realm in essence would imply, by definition, that he or she must be eternal and spatially omnipresent with respect to us, who are inside this box which is our universe.

Thus, a postulate here is that if change in universal space and time are solely dependent on the gravity and energy of the physical mass that make up our universe, then someone or something standing outside our universe must be independent of that space and time. Gravity and energy wouldn't affect the outside observer, so that the observer would be not only spatially omnipresent and eternal without beginning or ending, but the observer would also necessarily be omnipotent with respect to the universe. And if that observer is responsible for the creation of that energy and mass of the universe, then the omnipotent characteristic with respect to the universe would be proved by the existence of the universe itself. The other consequence implied by this postulate, however, is that the gravity and energy which gave rise to the universe also gives rise to any increase in universal space and time. This would be a controversial point to be sure. Some would argue that there was an infinite spatial vacuum, and perhaps even a negative time, during the period before creation (if we refer to that event as time $t=0$); some cosmologists model exactly this scenario when they speculate about no boundaries in the universe. But the postulate suggests that space and time are simply dimensions of universal reality that would have come into existence during the singularity of creation (the big bang); and thus space and time will obey the same laws of physics which those things that brought them into existence must obey, namely gravity and energy.

On the other hand, energy and gravity are not needed by space and time once they exist, while the vice versa would be required.

Now this argument may seem like a contradiction with the above postulate on first glance. However, the key word here is *change*. In order for space and time to grow or change, then energy and gravity are in fact needed, whereas they're not needed if space and time are frozen. These statements could be considered as fundamental boundary conditions for the singular point of creation itself, and could be a paradox if the concept of negative energy and gravity were not added to the picture. Let's consider two snowcones connected to each other at their sharp end, representing the origin of the universe, in a hyper dimensional space. Each snowcone would represent a universe, one growing in positive time and the other in negative time; each snowcone cross section, representing the physical space of each universe, keeps getting larger with time. Positive energy could be stored in one snowcone if negative energy is stored in the other, in order to maintain the conservation of energy, which was 0 at the origin; in other words, these snowcones together sum up to absolute zero in space, time, and energy. Yet individually, a universe is seen to grow and form. Hence the singularity of the origin of our universe could be understood at least from a geometric point of view if one considers a negative universe to offset ours. We'll see in a discussion on quantum mechanics, this is exactly what becomes possible, except that the negative universe is really considered as an imaginary universe in our universal frame of reference. And negative energy and gravity will come back into the picture during a discussion of wormhole travel later on.

Unfortunately, what physicists cannot begin to measure are the drivers, characters, and creators behind the singular initial spatial point of universal gravity and energy on either positive or negative time axis, since they obviously cannot be found or understood from scientific investigation or observation; but religious people will simply sum these up as being acts of God; and thus some features of God the Creator would be revealed. Cosmologists will simply call this point of singularity the infamous big bang and that its occurrence was a mere quirk of quantum mechanics; we'll talk about this in regard to quantum mechanics in more detail later. And as for how space and time expanded from this initial point of singularity to form our present universe that are in agreement with observation, we have only quite recently and accidentally started to find theories that have gained any acceptance; and the year 1980 was the turning point. In the next section, we discuss these theories and how they came to be. But first, we need to understand how this idea of a big bang came about.

Amazingly enough, Einstein had no intention to derive a big bang from his General Theory of Relativity; yet it became one of its greatest features and implications. But he grudgingly accepted the notion that there must have been some sort of beginning to all this madness by what his equations were telling him, and their consistency with the established observations of the continuing expansion of the universe. It took Russian born scientist George Gamow (1904-1968) to actually come up with the infamous fiery ball scenario, dubbed sarcastically at the time as the big bang, a conceptual original singularity of nearly infinite density, temperature, and pressure. Primordial high-energy photons left over from the initial fireball were predicted by Gamow to be still existing today as cooled off very low energy radiation left over from the origin of the ever expanding universe; only problem was that no evidence for it had been discovered yet. But in 1965, along came Arno Penzias and Robert Wilson of Bell Telephone Laboratories; they were tuning a small horn antenna for doing radio astronomy measurements when they noticed a constant low level static noise disrupting their reception. Looking for problems with the equipment or other malfunctions, they started pointing the antenna in different directions with no luck getting rid of it; the blasted signal was still there. It turned out that the antenna they were using was extremely sensitive and powerful by design; so they stumbled upon the idea that it was a cosmological phenomenon. It was in fact a weak cosmic microwave background signal left over, they theorized, from the big bang nearly 14 billion years ago, and one that Gamow had proposed should have been in the universe all along. For their discovery, Penzias and Wilson received the 1978 Nobel Prize in physics. As further higher resolution studies have detailed the cosmic data better in recent years, some overly zealous physicists have suggested that looking at this data is like looking at the "face of God", although I'm here to tell you that it isn't even close! As we will see later, this higher resolution data has described a young universe still many thousands of years after creation; but it has been quite useful in helping to both refine and refute possible theories of creation.

As a sidelight, the story is told of Einstein and Gamow crossing the street one day on the campus at Princeton University, where Einstein was resident during the 1940s, when Gamow mentioned that a colleague pointed out to him that his theory of relativity also implied that the attractive (negative) gravitational energy in the universe should cancel out its total repulsive (positive) mass energy such that stars could be formed out of nothing. Einstein just stopped in the middle of the road and just froze to think about it, as cars were whizzing by in each direction. The thought was interesting, but it implied the big bang (without quantum mechanics) should have simply closed in on itself, like snuffing out a candle. Quantum mechanical effects hadn't been applied to the theory at the time, if for no other reason, Einstein steadfastly shied away from quantum mechanics whenever he could. And here is where Einstein may have made his biggest real mistake, his own admitted "blunder" (we'll discuss later) not withstanding. Had he accepted quantum mechanics early on, he could have incorporated quantum mechanics into his General Theory to explain how the big bang would have occurred without caving in on itself. As it was however, he just threw his hands in the air and moved on. Now one of my undergraduate physics professors under whom I studied was at Princeton around that time; so I suppose he would have had better stories to tell if Einstein had made that connection! As it is, this professor was the one who loved to write complicated formula derivations on the board, turn around, scowl at the class, take off those horn-rimmed glasses of his, and simply bellow out, "How are we doing?" in that deep sarcastic but serious tone of his.

Now one of problems that we have noticed with the cosmic microwave background that has baffled cosmologists for a long time is the extreme distributed uniformity of space. Particles are evenly distributed throughout the Cosmos; and measurements of the uniformity in the cosmic microwave background data in every possible direction had cosmologists scratching their heads. There was no reason to believe in conventional big bang theory that there should be a casual relationship of the distribution of particles and matter throughout the outer Cosmos; by this we mean up to 10^{26} meters away in any direction. (10^{26} meters is the distance anything with light speed could travel in the lifetime of universe.) Beyond this, there is only uniformity of space implied by the cosmic microwave background data. And then there was the issue of flatness. Cosmologists couldn't figure out why the universe seemed to have just the right density to continue its observed expansion in all directions without either wanting to pull itself apart (allowing no planets and stars to form) or wanting to collapse back in on itself due to gravitational impulses; the horizon simply appeared flat. Standard models of cosmology failed to explain any of it; and as we mentioned, Einstein himself disbelieved his own equations which told him otherwise, because Einstein being Einstein wanted to show why the universe should be flat and static as it seemed to appear at least.

Of course, it's hard to resist the theological implications of this debate. Scientists as a general rule don't like to talk about something they don't know; but it was very strange that there was this coordinated uniformity and flatness of space in all directions, if in fact there had been a common starting point, as the General Theory of Relativity and further observation of receding galaxies by Hubble and others had implied. Many theories started popping up, such as multiple big bangs that must have occurred in different sectors of space; but that led into areas scientists didn't want to pursue, namely, that there had to be an organized force behind it all. Nobody in science was going to simply suggest God just clicked his fingers without trying to understand why or how, certainly not Einstein! Theologians and theists certainly were happy with the idea however, because it implied that there was a Genesis of creation after all, which would lead back to a God in control of it all; even a singular big bang, implied by relativity, would due for them. Well, this debate would actually continue unresolved past Einstein's lifetime and into the late 20th century without any one good theory to explain these problems, when one unsuspecting and unwitting young postdoctoral associate in a different field of study all together would come along with a theory that did explain them, except that he was looking to find an answer to a particle physics problem that had nothing to do with these issues whatsoever. This theory is what we explore next; it was given a name, inflation. And as we will see, it has helped to explain many problems in cosmology, while at the same time it has created others. The one theory that inflation did not try to supplant, however, was Einstein's relativity, and a good thing too; the more experiments on relativity that have been done, the more it has gained widespread acceptance. The trick has been how to make these theories compatible; and that is where even newer problems started to surface, in turn leading to more clues about our creation, and creator. And those clues have everything to do with the universal expansion of space and time.

VI. Space and Time

Alan Guth was a theoretical particle physics postdoctoral fellow at Cornell University during the late 1970s who was studying the production of magnetic monopoles – a particle with an isolated south or north magnetic pole - in the early universe. The only problem was that no magnetic monopoles have ever been detected or observed in the universe, at least that anyone has ever directly seen anyhow. But Guth calculated using conventional elementary particle physics that the early universe should have produced an overabundance of them, in agreement with John Preskill's work at Harvard. So particle physicists had a problem; where were those missing magnetic monopoles? We have electrons on the electric side, and we have dipoles, both magnetic and electric; but there just seemed to be no magnetic monopoles though.

It turns out, as an aside, I was also a postdoctoral fellow at Cornell about 15 years later; but my field was *experimental* particle physics. I wasn't so brilliant to be looking for magnetic monopoles; my work was much more elementary (physicists, excuse the pun). In my case, we were helping Cornell build another small experiment at the Fermi National Accelerator Laboratory in Batavia, Illinois, to measure the total cross section for proton-antiproton scattering at very high energies in the center of mass (about 2 trillion electron volts, or thereabouts). Since my experiment was simply a matter of measuring how big a proton was seen through the "eyes" of an antiproton, and vice versa, it didn't touch much on the subject of magnetic monopoles; but no matter, nobody still had seen one. Of course, when you smash ionized hydrogen atoms at those kinds of energies, you'll be surprised how many protons end up being slaughtered into a whole range of by-products. But still, nobody in the experimental community has ever yet had a definite answer to the missing magnetic monopole problem, although many an experiment has tried to find them.

When Guth moved on to Stanford, however, he stuck with the problem – where were all those magnetic monopoles? He finally suggested that perhaps conventional physics isn't going to explain this, and constructed a model of the early universe that could explain it. He called it the inflationary universe. It would explain how the outer reaches of the universe came to be so uniform and why it began so close to a critical density that it kept the universe from collapsing on itself while at the same time allowing it to expand at just enough speed that still allowed stars to form. It would also explain what happened to all those magnetic monopoles. He proposed that the expansion of the universe was propelled by a *repulsive* gravitational force generated by an exotic form of matter, all of which occurred at 10^{-37} seconds after a singular initial big bang of creation. This time was important because the infinite density at the singularity point of creation needed enough time to allow matter and energy to "communicate" and interact before inflation could occur, and yet provide the uniform distribution that we see today in the outer cosmos in every direction. Then the universe exploded in size by 25 orders of magnitude at just 10^{-34} second later, after which the normal rate of expansion of the universe would continue at light speed to the present, thereby giving cosmologists today the illusion of flatness of the horizon, the notion that the universe is indeed so gigantic that the horizon appears flat. And as for those monopoles, well, they're hypothesized to still exist; but with all the universal expansion, they become so sparse that no experiment was simply going to find even one very easily. Thus the inflationary theory had an explanation to many problems in the fields of cosmology as well as particle physics, all at once, ironically by combining one field of study, cosmology (the study of the infinitely big), with another field of study, particle physics (the study of the infinitely small).

Inflation assumes that the initial universe started as a "false" vacuum. It was false because instead of being a true vacuum of nothingness, it was a vacuum filled with virtual particles and antiparticles popping into existence momentarily and annihilating each other. Thus there were no particles or matter or radiation in the vacuum except a non-zero energy state; and it was called the cosmological constant by the theory. The cosmological constant determines then whether gravity is a repulsive or attractive force on mass, and can be used to model the missing mass of the universe, for instance; it seems to be still missing by the way. Anyway, as the theory tries to explain, this vacuum, containing the early universe, expanded exponentially during the period of inflation, due to the non-zero energy state that it had; the vacuum's virtual density would not decrease as space grew in volume, since its non-zero energy state represents the energy of empty space itself! After the period of inflation, the universe would have been sufficiently cool for a energy state phase transition to occur. Energy perturbations could occur in the fabric of space and time that would cause the birth of galaxy clusters. Matter was held together by gravity and indeed combined to form greater densities of matter, thereby allowing stars and planets to form, which in turn formed galaxy clusters that all began to orbit each other. The cosmic background radiation we observe today thus arose out of the initial energy perturbations in this vacuum energy state of the early universe; and the universe today would be what has evolved from this initial vacuum energy state.

Guth's self-admitted problem with his original theory, however, was that it filled the universe with a mess of expanding clumps of pre-galactic energy/matter bubbles (yes, bubbles!) that would keep colliding with each other and go in all different directions, that we not only see no evidence for, but in fact observe a universe that is simply too homogeneous for this to have been the case. At an international meeting of scientists in Moscow, 1981, at the height of the Cold War no less, Stephen Hawking, a Britain, gave a paper where he suggested inflation simply didn't work. However, another cosmologist was there; his name was Andrei Linde of the Lebedev Physical Institute in the Soviet Union. Linde (now at Stanford University) managed to convince Hawking that if some modification to the theory was made, the bubbles' problem would go away and the theory could work. They called it "new inflation". Funny thing though, during the conference, it was Linde's job to translate Hawking's paper into Russian; but the paper challenged the very argument Linde was trying to make about inflation's feasibility in the first place, which created a bit of an embarrassing situation for both scientists. In any case, they eventually got together and came into agreement that Linde's inflation model had merit.

By 1982, inflation theory had caught wildfire in the cosmology community and had already gained acceptance by most cosmologists around the world; but it has since matured into something called "chaotic inflation". The theory suggests that our universe actually grew out of a quantum fluctuation in some pre-existing version of space and time; and here is where it gets really creepy. Equivalent scenarios exist in our own universe for creating wholly new regions of inflation, e.g. the creation of whole new baby universes from within our own. One way this could happen, for instance, is for a black hole to collapse into a singularity that creates a whole new set of space and time dimensions in the same way the big bang created our own. Thus the question of where we come from goes away because the chain of events that caused our universe to form were going on infinitely in the past. Remember that time itself would begin at each of these singularity points, including during our own big bang, for the respective universe that came into formation. Thus, Giordano Bruno's predicament of how to save multiple universes from condemnation would take on new theological importance in this scenario; and you have to believe that the theologians of the world not only would not want to take this idea seriously, they just as certainly would not want to deal with the predicament! For the philosophers of the universe, however, it might be a different story! But you have to ask the question: if all these black holes are proposed to exist in the universe, aren't they busy creating their own little universes?

Well before you completely throw up your hands at the whole idea, coincidental evidence has been uncovered that gives some credence to inflationary theory. The Cosmic Background Explorer (COBE) in 1992, and the more precise Wilkinson Microwave Anisotropy Probe (WMAP) in 2003 have measured the cosmic microwave background from the early universe with very high energy resolution (to one millionth of one degree Kelvin in the case of the WMAP) and have found that the cosmic microwave background is indeed remarkably uniform across the entire sky. The results also showed quantum energy fluctuations at very small angular scales, the fluctuations being greatest at one degree radian, as inflationary theory had predicted for a flat appearing universe. These quantum fluctuations are seen as faint ripples in the cosmic background radiation and were predicted by Guth and others as early as 1982. It turns out that some cosmologists today are even actively pursuing theories about what conditions would allow a new inflation of repulsive gravity to be ignited in a laboratory environment. That's right; the idea is how one would create a universe in a bottle. A whole new universe could be simulated which would immediately decouple orthogonally from our present universe through some sort of wormhole effect, and thus do no damage to our own; and I'm not kidding. It certainly cannot be done without quantum mechanics; but by making use of something called quantum tunneling described by a quantum theory of creation, it might be theoretically possible, so some physicists say. Needless to say, it hasn't been tried; but we'll discuss wormholes and other theories of quantum creation later.

Perhaps a cosmically aged advanced civilization desperate to escape a dying universe would take a more serious interest in such an experiment, however. Recall that during the desperate race to build the nuclear bomb in World War II, some scientists involved with the Manhattan Project, where the first bomb was designed and tested, worried about the feasibility of a nuclear chain reaction causing ignition and fusion of hydrogen in the atmosphere itself; after all, the splitting neutrons and intense heat had to go somewhere. And the high hydrogen content of the atmosphere would be just the ticket to take them in; scientists had to be sure those hot hydrogen atoms wouldn't have enough energy to fuse and create helium, thereby basically simulating a mini-sun on the earth! It was finally determined however that a high enough density could not be achieved for such an unpleasant side effect to occur; so the development work continued. Anyway, the same safety concerns would be raised in any real quantum universe creation experiment; and the scenarios for it to occur are very far fetched. As we shall see, a universe creation scenario depends on having enough extreme heat and pressure in unstable space and time; and that's not easily done. And that means the idea becomes more far fetched by the second, excuse the pun!

Then there is the dark matter question. This isn't something out of **Star Wars** you understand. Physicists still have some very real quandaries on their hands, if they ever decide to accept inflation theory fully. And it's too bad too; inflation explains a lot of things. But it doesn't do a good job explaining where all that missing mass, called dark matter, is in the universe, because it sweeps it up into a "cosmological constant"; and we'll see below where that came from. Now cosmologists seem to know that mass is missing because models of the universe's observed expansion and the rotation rate of galaxies about their center of mass suggest it is. And a lot of it seems to be missing too. Further, the General Theory of Relativity, which predicts what galaxies' rotation rates and universal expansion rates should be, and cosmology's inflationary theory, which predicts what the critical density of the universe should be in order for it to have a stable expansion, cannot agree on what percentage of mass is actually missing in order to describe what is observed. Thus, for these theories to match, dark matter that we so far can't see must come in different forms. Some of it is cold (fat massive particles that sit around all day circulating in their galaxies), some of it is hot (fast little particles racing across the universe at near light speed, meaning they're relativistic, but with some mass), and some of it isn't matter at all – it's supposed to be dark energy. Physicists have names for some of these hot and cold particles; but they're not as swift about the source of dark energy, except that they know that it must have a repelling effect on gravity. The cold particles are mostly called – and I'm not kidding - WIMPS (weakly interacting massive particles). The hot particles are neutrinos mostly. And then there are the MACHOS (Massive Compact Halo Objects), which mostly are big bad massive planets, neutron stars, brown dwarf stars, black holes, and ordinary baryonic matter; what you need to know about these is what they all have in common, namely they don't emit radiation that can be seen easily. Cosmologists think all that missing mass, along with what we can see, is about one third of the makeup of the universe, and the other two thirds is actually that dark energy we mentioned. As for why some is mass and some isn't, we'll get into that below. But the particle that is supposed to give mass to the universe, by the way, is a particle that some scientists like to call "God"; it interacts with the WIMPS, it interacts with the neutrinos (if they have mass), it interacts with itself, and it is, believe it or not, called a boson. You'll hear about that later; but you know, when there is a boson around which scientists like to call "God", then you know there's likely to be trouble.

Now inflation theory could actually accommodate relativity theory if it also allows the universe to accelerate its expansion as it cools (that's a physicists' nice way of saying that one day everything that holds us together will eventually fling apart); but then it has to make use of that cosmological constant, the fudge term that Einstein had considered and stuck into his relativity model, but he – along with everybody else – couldn't figure out why. Einstein called it his cosmological constant too, although I think he was first to use it; he also regretted ever using it, and considered it to be the biggest "blunder" of his career. Now you know why a guy is considered smart: when the confessed biggest mistake of Time Magazine's Person of the Century turns out to be the key factor in making two independent universal creation models agree with each other, then you know it's time to take notice of that person. Unfortunately for the rest of us, we never got to hear Einstein's opinion on the recently theorized and still heretofore unseen dark energy that many cosmologists have been starting to worry about since 1998; but don't worry yet, we're getting there.

The thing about this story, however, is that when Einstein first proposed using his famed cosmological constant, he was trying to make his equations produce a result that would predict that the universe would not collapse in upon itself and be static; the problem was that his equations were predicting that the universe probably would collapse if the constant wasn't there, according to the theory. A static universe is one that is neither expanding nor collapsing. So he introduced his constant, which amounted to simply nothing more than a mathematical fudge factor, into his General Theory of Relativity in order to prevent the ultimate collapse of the universe, in his equations at least. Years later, when Edwin Hubble observed that the universe was in fact expanding, and not collapsing or stagnant, Einstein did the proverbial Homer Simpson "Dhoa!". And that is why he said it was the biggest botch of his career (as if he needed one), because his equations had also suggested all along (if he hadn't fudged with them) that the universe could be expanding or collapsing. It turns out he was right the first time, but only if the expansion was at a linear rate; and that's what Hubble found out, at least for the nearby galaxies he could see with his telescope.

Now Einstein's fudge term was not really supposed to be there anyway because it physically suggested that the initial state vacuum energy of the universe (before the big bang) was nonzero; and that's not really supposed to happen, because we all know that God keeps good records, especially with his law of conservation of energy. But today, it turns out the tables might have just turned once more on Einstein (and everybody else), because recent supernovae data taken since 1998 of the outer cosmos, has suggested, among other things, that the universe is not only expanding, but is in fact accelerating its expanse. Whoa! Stay tuned for the impact of this result, because that's how the whole riddle with dark energy got started. But first, there are a few more riddles to talk about.

Enter John Anderson, NASA physicist at the Jet Propulsion Laboratory in Pasadena California, whose job it was to monitor the Pioneer 10 and 11 spacecraft launched by NASA in 1972 and 1973, respectively. He noticed something quirky about the measured yearly distances that the spacecraft were traveling versus what their expected distances were supposed to be; and the discrepancy appeared to be the same each year he did the comparison. There was a strange

13,000 kilometer deviation in the crafts' expected position traveling over 352 million kilometers per year. And the effect has continued till the last we heard from the Pioneers in 1995 and 2003; now they have sped past pluto and our solar system into extrasolar space. A negative acceleration constant factor of 8.7 * 10^{-10} meters/second^2 was measured opposite its traveling path direction for each year; and that's the same discrepancy factor that cosmologists say that galaxies have rotating about their axes, which gave rise to the concept of dark matter in the first place! Funny thing is that no one has yet been able to find a suitable explanation for the Pioneer phenomenon because while enough dark matter could be contained inside galaxies to cause this effect, certainly not enough space between the Pioneers and earth exists for dark matter to cause this effect. Furthermore, data from other craft (the Galileo and Ulysses) are also now bringing back inconclusive but consistent results with the effect. Relativity doesn't explain it (at 40,000 kilometers per hour, that would be a few football field length effect per year – but be careful, I calculated it), solar radiation and solar wind doesn't explain it, and gravitational effects of the planets, comets, and the sun don't explain it. So what does?

To explain it requires consistency with other observations that we do measure with great accuracy, for instances the orbits of our own planet and moon, if for instance it was due to the sun or earth in some way. Other types of new physics and resistance from dark matter run into theoretical problems with the models of expansion of the total universe. Later we will discuss other possibilities for this result; but the only conclusion that is certain so far is that the effect has something to do with the fabric of the universe itself. How else would the measured discrepancy of the acceleration factor for the Pioneer craft be the same as that for distant rotating galaxies?! The effect has come to be known as the Pioneer anomaly.

In a published review by Anderson's team (Anderson et al., ***Study of the anomalous acceleration of Pioneer 10 and 11***, Phys. Rev. D65, 082004, 2002), they stated that an anomalous constant acceleration is observed and isotrophically directed toward the sun. All internal systematic effects have been accounted for in their analysis. The effect has the implication that the universe appears euclidian, homogeneous, static, and seemingly infinite, something that inflation, as it turns out, effectively predicts due to the horrific expansion of the universe since the earliest moments of creation. But every point in space seems to act like the center of its own sphere of radius c/H_0, where c is the speed of light and H_0 is the Hubble constant, which arises in General Relativity as a proportional constant times the distance of a star system from earth to give its recessional speed going away from the earth, thus causing the Doppler redshift of light (as it loses energy from a receding source) coming from the star in question. The thing about this that gets to be quite interesting is that this heretofore unexplained backward anomalous acceleration of 8.7 $* 10^{-10}$ meter/second^2 of the Pioneer spaceships is strikingly close to the factor cH_0, the same unexplained acceleration that galaxies seem to have toward their center, which the dark matter was designed to explain. Further, this unexplained percentage shift in acceleration measured year over year is about the same percentage change as the percent change in the age of the universe year over year. Could some leftover ripple effect of inflation on space and time itself be causing it? Could it be a phenomenon of the heretofore unseen dark energy or even vacuum ground state energy of the universe; and if so, what are the values of these energy terms anyway? Physicists have been prepared for a long time to accept that the Hubble constant is meant to be a measure of the resulting linear spatial expansion abnormality in the universe; but it would need new physics to explain what else it might need to absorb.

Then in 1998, teams of astronomers from the High-z Supernova Search Team and the Supernova Cosmology Project, surveying the sky for the most aged type 1a supernovae in the universe and spread across the horizon throughout distant points of the galaxy, made a jaw dropping discovery. Supernovae were chosen because of their relative brightness in the sky; they are the product of thermonuclear explosions of white dwarf stars that are 40% more massive than the sun but with a radius 100 times smaller. Given models of inflation that suggested the universe was supposed to be slowing down in its outermost expansion, cosmologists theorized that they could accurately measure the rate of this slowing expansion by using the most aged (10 billion year old) type 1a supernovae that could ever be seen. And according to the General Theory of Relativity (which has held up to this time), the measured redshift of light from the stars was in proportion to the distance that the light source itself was away from earth. What the astronomers found (Riess, et al., ***Observational Evidence from Supernovae for an Accelerating Universe and a Cosmological Constant***,The Astronomical Journal, V116, p.1009-1038, 1998) was that the supernovae were not only not slowing down at all, they were in fact speeding up! They were doing exactly the opposite of what all that theorized dark matter needed by inflation theory was saying they should do! The expansion was creating a repulsive effect on the gravity in the universe; astronomers almost euphemistically referred to its explanation as due to something called, and yes, here it is, that ubiquitous dark energy. The Hubble constant had no real way of handling this phenomenon. And later in 2003, more recent data from the Wilkinson Microwave Anisotrophy Probe, the Two-Degree Field Project, and the Sloan Digital Sky Survey Project have provided more evidence for the same effect; and moreover they have shown that overdense localized regions of space, that would normally experience a predictable gravitational collapse, have had their impending implosions slowing down, now known as the Sachs-Wolfe effect.

What this all means is that physicists, specifically cosmologists, have a problem. They need dark matter to provide attractive gravitational energy to explain the fast rotation rates of nearby galaxies, which would inhibit universal expansion rates, while at the same time they need dark energy to explain the repulsive gravitational accelerated expansion of the rest of the cosmos. What's a scientist to do? Be assured, however, that when someone starts talking about something that is dark, as in it can't be seen, there's a good chance that he or she may not know exactly what it is he or she is really talking about!

Well, curvature of space and time in the outer universe could explain some of these effects, except that very high resolution data from the WMAP data of the primordial cosmic ray microwave background has not noticed it inside the visible universe; that data has tracked tired photons that have traveled the cosmos since about 380,000 years after the big bang. In fact, the data taken back to that age has shown the universe to appear quite flat and Euclidean, in agreement with the Pioneer anomaly implications and inflation theory. Curvature has only been observed locally to warp space and time when light passes through localized gravitational wells of massive stars, as predicted by General Relativity. Inflationary spatial ripple effects in the early universe would result in different anomalous accelerations of the Pioneers versus the galaxy rotation rates, and not the same as is observed. And as for dark energy, we would have to know something about dark energy. Trouble is, cosmologists don't know a whole lot about dark energy; and their explanations of its source get more confusing from there. However, we can digress for a moment about what they think it is.

Remember that cosmological constant introduced by Einstein? Well, it turns out that cosmologists were only too tempted to take another look at this "fudge factor" to see what was in it for them. Problem here is that the constant can only explain a nonzero vacuum state energy of the initial universe, and models need a value of the order 10^{120} for it to add up to the dark energy needed to produce the massive effect that is seen with the supernovae data. Wow! Then there is something called quintessence; this is a dynamic energy field of negative pressure due to quantum particle excitations of mass 10^{-33} electron volts. At least quintessence is a plausible explanation for the onset of cosmic acceleration during the big bang, as we will see; but using something that is 10^{-33} electron volts to explain today's cosmos outer expansion also seems far fetched! And finally there is the idea that dark energy is just masking the possibility that General Relativity theory needs modification in order to account for this extraordinary accelerated expansion of the outer cosmos.

Perhaps none of these theories about dark energy seem very realistic to explain today's cosmological problems. To the outside observer, it may seem like a lot of gobbly-gook to create more fudge on top of the fudge already in place surrounding dark matter. And believe me; it's really uncharacteristic of a physicist to do that. On the other hand, it shows really how desperate the stage still seems to be set in cosmology. Remember that cosmologists still don't have a competitive theory for inflation, which even shows some incompatibility with relativity theory and observation over the dark matter question. And speaking of which, it would certainly be helpful to know what this dark matter really consists of, since we apparently can only see about 4% of the mass or energy of the universe anyway, namely ourselves, the stars and the planets, where the bulk of it, by the way, just happens to be only hydrogen and helium! Then there is the matter (excuse the pun) of the Pioneer anomaly, which has truly everyone baffled! Maybe there really is something wrong with all that heretofore unseen and unknown yet theorized dark matter and dark energy. It seems we are really missing something in the physics; and uh-oh, that usually means new physics. Perhaps it has something to do with space and time itself. Perhaps it has something to with multiple dimensions that might exist and are used by string theorists. Perhaps it has something to do with the speed of light, or other constants of the universe. And this is where the new physics just might take an unexpected turn. The constraints of any new physics however would be that the model has to incorporate the observed data from our universe already, where the dark matter and dark energy have so far gone missing.

Scientists, like stockbrokers, can easily get caught following the crowd; it's called human nature, and we all seem to do it at times. The only problem with all of this is that we are all human beings, regardless of our field; and we never want to be left behind. This is a natural bias of human nature. You'll be surprised how many experiments seek to confirm or find a way to confirm the best selling theories, while other ideas are systematically ignored. In religion, you find the same phenomenon; and do you ever! When Christians recite their creeds, they can't get enough of the God-in-three-person Trinity; yet the earliest Christians, including Jesus, St. Paul, and the Apostles didn't know what it was! It would be simply delicious, in fact, to hear what St. Paul would have said about that; but he did warn about the pagan myths that would surface from time to time! And pity those poor theologians who are required to stick with what they have been taught for fear of the threat of admonishment or punishment from those on high if they so happened to teach what they really thought – and I don't mean punishment coming from the outer realm!

In any case, there is a real problem with the creation story, and I'm not referring to Genesis. Physicists and cosmologists know they have a problem fitting the universe together into a model that meets all the existing data; and there is nothing new about that, except that a lot of time and effort is going into efforts to find, and even circumstantial efforts to find all that dark matter and dark energy. And of course, circumstantial evidence is usually another way of saying the real evidence hasn't been found yet. What happens if they're really wrong? What happens if there really is new physics needed to explain what's going on? To their credit, there are distinguished physicists out there who have started to ask these questions, and others who will strive for more concise data on the measurements taken already. As for the constants of the universe, namely the speed of light, Newton's gravitational constant, the Hubble constant, the Planck constant, and so on, there does seem to be something we are clearly missing. Perhaps these constants aren't so constant after all; maybe there is some dependence with the fabric of space and time. We know, for instance, that there is a relationship between gravity and the fabric of space and time, and gravity with mass. What about the dependence of mass on the fabric of space and time for instance? Perhaps part of the problem lies there. But to detect new physics, and get people interested in it, without the scoffers showing up to drown you down with laughter; it won't be easy. In this situation, it's the pundants of science who prefer to ruin someone's reputation rather than make real progress of understanding; and they take on the parallel roles of those religious officials in charge of excommunication. And herein lays a real problem in finding new physics, or updating theological doctrine, for that matter. No one wants to be the fool with a theory that no one appreciates.

If this kind of problem somehow seems familiar, it should be. Copernicus faced the exact same situation among the intelligentsia and powers to be. He saw that the astronomers were quite happy at the time with the geocentric notion that the earth was the center of everything. After all, the stars seen from earth had a reproducible and predictable pattern that fit with the medieval calendars just fine year after year. The church had taken its stand that humans must be the focal point of the cosmos as far as God was concerned; and by extraction, the earth must be the center of where "it's at", wherever that was. As more data was becoming available during the time, postulates and adjustments to the existing theory were made, all with the church's blessing – just as long the earth still stayed at the center of things. Finally, it took an astute mind and convincing data to throw out the "baby with the bathwater" when it became clear that the baby wasn't a baby after all. Unfortunately for Copernicus, and mostly for Galileo, the church's acceptance would still be required; and it would be centuries in coming, as with anything in theology!

One alternative of presently accepted theories, for instance, could be to use a geometric construct of the universe that makes it a subset of a higher dimensional hyperbolic finite superspace. In this case, our limited dimensioned universe would in fact look infinite because light could travel unlimited distances by wrapping around the universe such that the universe would still look flat and uniform; and the mass density of the universe could be accounted for. Only real problem that doesn't go away is that we should be able to see repeatable characteristic patterns in the cosmic microwave background, after light has had enough time to reach the cosmos outer boundary, which so far is not seen with the 2003 WMAP data on the cosmic microwave background. According to inflationary theory however, the outer bound of the universe cannot be reached by light coming back to earth for another 10^{25} years; so much for seeing the universe wrap around in this life time! But this idea at least gets closer to the idea that God belongs to a larger hyper dimensional realm inside of which our fishbowl universe exists, except that God would also be confined to his own higher dimensional superspace fishbowl as it were. If he is to be continuous and omnipotent, his realm should not be confined to any kind of fishbowl, no matter how many dimensions he has! String theory, with its inclusion of other hyper dimensions, is one of a family of theories within the very premature subject of quantum gravity that is getting a lot of attention these days and seeks as its primary motivation is to find the ultimate Theory of Everything by combining a description of gravity with all the other forces of nature; it encompasses this idea that we live in a small subspace of a much larger hyper dimensional space that allows the hyper universe to regain its symmetry by recombining all forces of nature into one unified force. And this force, as we'll discuss later, bumps physicists up against their own uncomfortable description of God within his own supersymmetric realm.

Other interesting ideas that keep theorists humming in their cubes have ways to fold, bend, and turn inside and outside existing General Relativity and inflation theories in order to find perturbations or asymptotic values within more conservative mathematical constructs that allow existing theories to agree with the observed data. This is always a neat approach in that it avoids new physics, keeps one from being laughed at, and tries to explain things from that which has already been accepted and understood. For instance, an intriguing research paper by Edward Kolb of Fermilab (Batavia, Illinois, USA), Sabino Matarrese of the University of Padova (Italy), Alessio Notari of McGill University (Montreal, Canada)) and Antonio Riotto of Istituto Nazionale di Fisica Nucleare (INFN) in Padova, Italy, (*Effect of inhomogeneities on the expansion rate of the universe,* PRD 71, 023524, 2005) shows that by adding an additional universal Hubble parameter to the normal Hubble constant (designated as H_0), one can account for "super" Hubble perturbations that are due to supposed inflationary long wavelength ripples that may have been created during inflation in the primordial space/time universe. The effect, when overlaid onto the local universe, would cause the outer universe to appear effectively as expanding in an accelerating fashion, according to a local observer inside the local universe where the ripple is not apparent.

The time dependence of the Hubble constant is a hotly debated topic since its value becomes the inverse of the age of the universe itself, within the confines of General Relativity theory; in relativity, it is defined as a prominent proportionality constant for measuring the linear expansion of the universe: distance from earth times the Hubble constant gives the recessional velocity of an object moving away from earth. We'll see that theorists who ponder a universal Theory of Everything, however, may find that it needs more generalization to include the outer boundaries of the universe, since its measurement has been distinctly tied to more local objects in the universe, thus the justification for playing with the Hubble constant in the paper mentioned here. But the paper also assumes cold dark matter still exists inside the local universe, which is yet to be found; and it doesn't try to shed light on the Pioneer anomaly. Yet the idea of the existence of primordial ripples in the space/time fabric of the universe is an interesting possibility, to be sure; and it removes that irritating cosmological constant from the equation that cosmologists keep reaching back for in order to explain the outer universe's accelerated expansion.

As for time itself, there is always resistance among scientists to consider known constants to be changing with time, the speed of light for instance. However, the Hubble constant is defined by time, and it is directly linked with other gravitational constant of the universe; so it shouldn't be such a surprising idea. Later, when we discuss light's role in finding ultimate symmetry of the universe, we speculate how the fabric of time in the outer edges of the universe may have something to do with a much faster speed of light in the early universe, along with the seeming accelerated expansion of the outer universe. Division of time itself could be responsible for the repulsive gravitational spark that gave rise to inflation of the universe itself; and it also gives rise to the intriguing possibility of a parallel negative universe that goes backward in time. If this seems ridiculous, consider that particle physicists quite naturally make use of a "time" operator to evaluate the symmetry of a particle interaction under time reversal, a subject that we shall see plays a major role (see section The CPT Conundrum) in answering why our universe exists and didn't happen to annihilate itself in the first place. And, by the way, did we mention that time only seems to go in one direction in our present universe? Forget all those **Star Trek** episodes, as interesting as they are, there isn't anyone who is ever going to wrap themselves around the gravitational field of the sun or any other star in order to produce a reversal of time. So what's going on here? Is it really a bad idea to build that dream time machine that could let you go back in time to bet on those race horses whom you already know have won?

Well, Stephen Hawking had the idea that by studying black holes, one could obtain a model explanation of creation's origin running in reverse. And don't look now, but some scientists think our own Milky Way galaxy, that includes our beloved earth, may be rotating right now about a giant black hole at its center. Black holes have the feature that gravity causes the collapse of matter into a theoretical singularity, whereas big bang theory has the feature that gravity was at least momentarily repulsive allowing matter to spew throughout the cosmos. The interesting idea here however is that, according to General Relativity, time itself is affected by the gravity well of the black hole. If one is situated at a reference point far away from the black hole, one would see time grinding to a halt inside the black hole, according to theory; similarly at creation, there would have been no time for us to talk about, excuse the pun. But here is where it gets scary; time itself could not have been conserved if the mass of the universe was to be born and expand like it did during inflation. Mind you, it didn't last long, sure enough; and we are talking about quantum mechanical effects here you understand, but we're coming to that. And speaking of which, Hawking proposed that a kind of radiation can emerge from the black hole that allows us to study it, called (you might have guessed) Hawking radiation. It is due to particle - antiparticle pairs that should annihilate, but sometimes don't because one particle manages to jump out of the gravity well of the black hole while the other one doesn't. Thus, as it is with black holes, it is the quantum mechanics of nature that also allowed the universe to escape its own vacuum existence, or lack thereof, from the very start. Only, Genesis seems to have left out that part in its version of the creation story; one would guess that either God couldn't get the concept through to the writer or the writer couldn't get the concept through to the reader, or whatever. What we can say is that the dawn of mass, gravity, and quantum mechanics itself in the universe had everything to do with the rise of space and time, and not the other way around! God must have had fun doing it, because it just goes to show that he not only did not need space and time for his own existence; but he probably used his own hyper dimensional toolbox to create them! And in doing so, his own omnipresence, eternity, and omnipotence with respect to our universe was assured.

From Newton's laws of gravitation to Einstein's General Theory of Relativity, we have learned a lot about the nature of space and time that constitute our universe. Something out there is holding our universe together as it continuously expands, and the fabric of space and time is intimately connected to the mass density and gravity in the cosmos. Inflationary theories explain fairly well how the observed universe evolved from the moment of creation; but it also has problems adjusting to predictions from General Relativity theory, such as how much dark matter and dark energy are in the universe. And where is all this dark matter and dark energy anyway that we can't see? Only badly contrived theoretical mixtures of dark matter and dark energy can accommodate the still existing problems in cosmology today: faster than expected galaxy rotation rates, the Pioneer anomaly, and the outer accelerated expansion of the universe. Inflation, if true, may have created its own ripples in the space-time continuum that no one has understood very well yet; but it does explain the flatness and uniformity of the mass distribution of the visible universe that we do see. General Relativity accurately describes the local expansion of the universe on a large scale, but fails to explain observed galaxy rotation rates without the inclusion of dark matter or some anomalous acceleration factor and lacks the ability to solve problems at the smallest scale, namely underneath the atom itself; and that's where quantum mechanics comes into the picture. And it is from underneath the atom, and indeed underneath the fundamental particles that make up the atom itself, where the asymmetry that gave rise to the mass of entire universe, and indeed to the big bang itself, can be unlocked and understood. Just as similarly, all the forces and fields of nature, which allow fundamental particles to exist and interact, hold the mystery behind the lost symmetry of the universe, and how it could be found or restored. In order to understand how to get there, we don't need to throw up our hands, like the Egyptian philosopher Valentinus, and say it was all God's fault. From studying and learning how to unify these forces and interactions at the quantum level, God has provided clues and left open the door for us to understand how he works, who he is, and where he comes from. And one clue, as we will see, is how come the anomaly in the Pioneer and galaxy rotation accelerations from what is expected in General Relativity are the same! The anomaly is very close to cH_0, where c is the speed of light and H_0 is the Hubble constant.

As we come to a gradual understanding of God's existence, and our place in it from our little corner and perspective of space and time, we can try to understand what eternity and omnipotence actually means. From there, an understanding of what God used to create the universe can be gained by understanding his favorite tool for creating the universe, namely quantum mechanics. And perhaps, just perhaps, this tool allows us to find the road to his perfect and symmetric realm; at the very least, our limited understanding of it has already accounted for most of our present day GDP! But that road that leads us to his total creation gets curvy, and even branches off at times with forks, leading us where we do not expect or even where we do not wish to go. And this is where the physics gets fun, and the joy of learning about it reveals just how large, and small, this chasm between his realm and our realm really is. God didn't likely intend for the universe to be a trivial creation, yet one with many strange, awe inspiring, majestic, and even unexpected turns and observations. But this is just another way of saying God seems full of surprises, and he probably wanted that way. There is a way we can understand him better by looking across that chasm; but he's probably going to make us work for it. When we do, we'll probably find that a very imperfect and asymmetric universe is dangling from a much more superior, perfect, and symmetric realm. At first thought, that idea may seem illogical; but whoever said that the logic of God was equal to that of man. Just expect to be surprised.

VII. Alpha and Omega

Who ever said "I am the Alpha and the Omega" must have known what they were talking about; and it probably was something more than knowing the first and last words of the Greek alphabet. Perhaps St. Peter in his second book in the biblical New Testament was on to something when he said that as far as God is concerned, a thousand years are as one day, and one day as a thousand years. Actually though, I think Peter was off by about 10 billion years or so. Now we can see why. The sun and earth alone have been around nearly 5 billion years; that's almost nine billion years after the creation of the universe. It becomes amusing then that people still get wrapped up over the seven day creation story in Genesis when we know now that one day, as we define it, did not exist on anybody's calendar during those first nine billion years that the universe existed. And then there's the small problem that time itself simply did not exist at the start (or should I say before the start). First of all, God had to create it. And what is time anyway, other than simply another measurable dimension of our universe that was started by the big bang. Mind you, one can get terribly bored in a realm where no timeline exists; and so I think God must have had problems with boredom too. The biblical story however can stand on its own if it is to be understood as a figure of reference for relative events in time as God sought fit to create them. Timelines are relative to a God because he is independent of time. Anyway, an Omnipotent Creator wouldn't have it any other way!

However, there is trouble in conventional Christian doctrine, at least, because it teaches that the Omnipotent Creator who formed our physical universe is also a Holy Trinity, namely a discrete being. The problem with that philosophy is that discreteness runs contrary, and even opposed, to the idea of continuity; and continuity is not only implied, but becomes a direct feature of omnipotence. Someone really needs to teach these theologians some quantum mechanics! In other words, the omnipotence of God requires that he must be continuous and not discreet. A postulate here can be made that the concept of a Binary Deity or Trinitarian Deity or even a Quad Squad implies a discreet quality that is inconsistent and contrary to the concept of an Omnipotent Being. An Omnipotent Being must be able to exist in a continuum of his creation and have no discreetness or imperfect asymmetry at all, if for no other reason, he must by definition be all powerful, and thereby have the ability to be all knowing, all encompassing, and all pervasive of his creation. Finite numbers imply concrete boundaries; and discreet entities go against the concept of omnipotence. However, most Christians believe in a Trinitarian yet Omnipotent God. How did they get into this pickle? For the answer, one cannot consult the Bible, because the Bible says nothing about the "Trinity" at all. For the real answer, let's digress to the year 325 A.D. to a place called Nicea.

The church had just been formally accepted by the Roman government with the "Edict of Toleration" put forth by the Emperor Constantine in 313 A.D. Up to this time the church had undergone a horrific persecution by Roman authorities for more than 200 years. But now peace and public discussion were beginning to flourish in the church and with it, unfortunately, divisions and arguments over church doctrine. To quell the various "heresies" that were being taught, Constantine himself convened a group of 318 "bishops" to the Council of Nicea for the purpose of hammering out a common doctrine; the result was the Nicene Creed, and the concept of God as "Trinity" was hatched. Now the church at this time had become much more centralized, and even controlled by the government in Rome, with the bishop of Rome given the most power, splendor, and wealth. As a result, the Nicene Creed was formally adopted and approved by the Roman government as a canon of the church. Constantine and the church bishops then tried to make the Christian faith as appealing as possible to the general Greek and Roman populace, who had grown up with pagan beliefs and Greek philosophy. So it's not surprising that churches were being built as great temples of architecture with many adoring relics and statues.

Well, this is all very nice. But the problem was that pagan philosophy was still creeping its way into church doctrine. In referring to God, Constantine and the church bishops were bouncing words around like "persons", "substance", "essence", "seen", "unseen", and "incomprehensible", some of which came directly from Platonic philosophy which deals with a person's oneness with the world that in turn is supposed to have its own soul or being; so it's not surprising that when the Nicene Creed authors considered God, they added that the Son is "of one being" or substance with the Father. Furthermore, Greco-Roman philosophy considered the concepts of Mind and Body as going in separate directions in relation to the ideal soul, whereas the Bible clearly kept them together as one. Unfortunately, the Christian philosophers of this era had recreated this distinction again. When one applies this dual image of the human being which includes the soul to the concept of God, one conveniently sees how the Trinity comes about. It turns out that our second century Gnostic mythology friend, none other than Valentinus himself, may have been the first to devise the notion of God as being separated into three subsistent hypostases or persons (Father, Son, and Holy Spirit) in his writings, as suggested in Bentley Layton's **The Gnostic Scriptures** (Doubleday, 1987). And by the time we get to the third and fourth centuries, Platonic thought was becoming rampant in some Christian communities. The New Catholic Encyclopedia even admits "From the middle of the fourth century onward, however, Christian thought was strongly influenced by Neo-platonic philosophy and mysticism." Lutheran doctrinal historian Johann Von Mosheim (1694-1755) complained that the church bishops of this era clearly based their thoughts on "Platonic philosophy" and then "extended and embellished" them. About the Council of Nicea deliberations, he goes on to complain, "There is so little clearness and discrimination in these discussions, that they seem to rend the one God into three Gods." Also Albrecht Ritschl (1822-1889), an academic German theologian, observed that the Trinity that came out of this Council corrupted the early authentic Christian message by introducing a "layer of metaphysical concepts, derived from the natural philosophy of the Greeks." And devoutly religious Sir Isaac Newton, of all people, spared no flattery when he did his own investigation of the history of the Trinity; he came to the conclusion that it was a pagan corruption and a fraud perpetuated by the early church and originated at Nicea by one Athanasius, a rather aggressive figure who was bishop of Alexandria, a city heavily influenced by Platonic thought, and from where none other than both Ptolemy and Valentinus had once hailed; Ptolemy was the guy, you'll recall, who thought the earth was the physical center of the universe, a theory the church collectively wrapped its arms around for 1600 years as a theological truth.

Now Athanasius himself was later exiled on three separate occasions by an increasingly politically polarized church, including once by Constantine himself after Constantine had also changed his own mind about the idea of the Trinity. No less than fifty years of strife ripped throughout the church over this one issue; and it persisted until Athanasius managed to gain the support of a clear majority of church officials over his main rival Arius, another aggressive figure who pushed the equally gallish idea that God the Father was a superior deity reigning over the Son and Holy Spirit as lesser deities. However, the seeds for the later separation of the Roman and Eastern Orthodox churches had already been sown. When western Christian authorities continued to insist that the Nicene Creed declare that the Holy Spirit "proceeds from the Father *and the Son*" (as if the Holy Spirit were subordinate to a dual persona), that did it; the two churches formally split in 1054 with a series of excommunication decrees. Too bad though, although both churches accept the Trinity (but in different ways), western Christians could take a page or two out of Eastern Orthodox theology, which asserts that no finite human can even begin to understand God as he actually is, because he is infinite! With that, I can agree!

So then, what was the result of all this shenanigans? Well, for one thing, Christian doctrine proponents and theologians throughout history have been left with a problem. They have wrapped themselves around a concept of God beyond the teachings of the Bible and the brainchild of a well-versed Greek philosopher, who was himself exiled by the church itself on three separate occasions. And religious theology, being religious theology, you just don't back off of a 1700 year old idea very easily, especially once it has been accepted as dogma. And if you ridicule it, you get labeled a heretic. Goodness knows what Galileo has to say about all of that. If the church had wished to go back in time and resolve this mess, it could have simply stated that the truly Christian aspects of Father, Son, and Holy Spirit are all part of God's continuing revelation of himself, the Omnipotent Creator, and not in relationship to and separate from each other, as if they were distinct and discreet persona. For Christians then, the divinity of Jesus is by definition of the omnipotence of God, and not by implication of the dual persona of the Greeks. But the church today still bends itself into a pretzel over the Trinity because it continues to make these Greek philosophical distinctions about God, while at the same time trying to say, well don't worry about it if they don't make sense because they are all really one in the same anyway. It's a kind of "have your cake and eat it too" line of argument. When God had revealed himself in the burning bush, he probably didn't mean for us to set it apart as an item of worship, unless he wanted us to burn down the house!

As a person of faith on an individual level, you don't really need to whip up a frenzy about all of this, however. If the church borrowed from pagan philosophy to write one of its cannons, well stuff happens! What you believe on a personal level, however, is much more important. And it helps to start with truth, faith, purity, and honesty with oneself. Jesus once said to those near him, if you don't believe in me, then believe in the things that I say and do. When you look and strive for love (feeling), truth (the word), and fellowship (common consciousness) in the world as ways in which God reveals himself in the universe and communicates with the world, as Jesus himself did, then you will know that you have found God. Yet you never hear anybody trying relate these as if they were three, and only three, discreet entities of one equal substance, as if anyone knew what that substance was supposed to be. St. Paul in his book to the Romans of the New Testament made it absolutely clear that "God is one"; he did not say "God is Trinity". Not once. Again, in his letter to the Ephesians he states clearly that there is "one God and Father of us all, who is above all and through all and in all." Never wishing to be one to argue with Paul, I could not have stated it better. We'll see later that there are concepts now being developed in string theory, a theory that tries to put all the laws of physics together into one Theory of Everything, which might find Paul's words to have been more prophetic than maybe even he had imagined; and with an all-encompassing God in the universe, we'll be able to speculate on the interface between God and our conscious.

Okay, so why all this fuss over the Trinity? Well, for one thing, if we were to have a real Theory of Everything, it would probably need a Grand Unification Theory of Religion as well! A singular perfect symmetric Omnipotent Creator at the center would be consistent with this approach, not that many theologians are trying very hard to do that. The second thing is that when one of the most formidable minds of all time has something to say about theology, after doing a very serious investigation of its present flaw, maybe someone should listen; in this case, we're talking about Sir Isaac Newton himself. Besides bringing out the physicist in all of us, a kind of GUT theory for religion would at least have the attraction that it would surely solve a lot of the present day world's problems and wars, and probably those of the past and future as well. But we can be sure that the church is in no mood to start another 50 year internal war over it, and certainly wouldn't do it just because the non-theologian Newton said so. But as an amusing academic aside, how would such a unified theory of religion work?

Well, Jews and Christians would probably have to stop arguing over the Messiah for one thing, although a starting point would be that Christians say he came in the person of Jesus and will come again, while Jews will say the Messiah is yet to come, but coming indeed. Divisions of essence, substance, and discreetness of the

Omnipotent Creator would have to disappear; and the church would need to purify itself from the infection and influence of Greek philosophy, such as with the Trinity. Muslims certainly also believe that God is one, unique, Omnipotent Being who rules over us and demands our obedience. Muslims, Jews, and Christians might even actually find themselves in common agreement with the one notion that every person of faith seems to agree on, that there is one absolute, omnipotent, and unique God, the creator and center of our universe, symmetrical and perfect in every way. As for what some physicists think, that might be a different matter, a different particle actually; but we're coming to that.

One problem people of faith need to resolve however is just who was who among the prophets and representatives of God anyway! History certainly seems to lean on the side of a GUT theory of religion, although like the universe itself, a symmetry breaking occurred that gave way to the world's major religions. We start from the prophet Abraham, who is the human father of religions to seemingly everyone. His wife Sarah bore a son Isaac who became the seed for the whole Jewish nation while his other wife Hagar gave birth to Ishmael, the father of Arabs who settled in Mecca and from whom the prophet Mohammad descended. And Jesus was the Christian Son of God who rose up from the Jewish nation and was the Muslim Prophet who visited India to reinvigorate Buddhism. (Whatever your beliefs, you do get the feeling there is something special about this man.) And then there is that interesting side story in Genesis where Isaac's wife Rebecca gave birth to two sons, Esau and Jacob, who would end up becoming founders of two warring nations, Edom and Israel respectively. Edom it turns out was a nation below the Dead Sea arising out of the modern state of Jordan. Evidence for its existence now dates back to as early as the 10th century B.C., during the time of Solomon and David, as found by an archeological team led by the University of California at San Diego and published in 2004 (Levy, et al., **Reassessing the chronology of Biblical Edom, new excavations of 14C dates from Khirbat en-Nahas (Jordan)**, Antiquity 302, December 2004). The Edomites of biblical history were subdued by King David; yet there are many West Bank Palestinians today who gain their heritage from them.

If these founders of faith (e.g. Abraham, Jesus, Mohammad) and founders of nations (e.g. Isaac, Esau, Jacob, and Ishmael) were ordained by the hand of God, then we can see that, along with the physical symmetry of the universe, the symmetry of many cultures and religions was meant to be broken along the way; and today we see the divisions among races and creeds around the world. We could philosophize (or even prophesy!) all day and all night why God broke up his own symmetry of *everything*; but it could well be argued that God had a real reason for doing it. After all, a perfect Omnipotent Being doesn't take action without a plan! We could take the Egyptian philosopher Valentinus' own naive view that it was God's own fault why we have this mess; but somehow I think Valentinus was missing something. There is a key that unlocks the door to solve the asymmetric problems that seemingly came from God's hand; and he must have wanted us to find it; otherwise we would have just been fodder for the dinosaurs and extinguished by his hand long ago. And that key involves learning, reasoning, and finding answers through our observation of the universe, "for then we would know the mind of God", as Stephen Hawking would say.

As for physicists, there is an almost fanatical attempt these days to find the lost symmetry in the universe; if they discover it, they would be well on their way to having a Theory of Everything. So alternate realities, even those that might be spiritual (although physicists wouldn't likely admit it) are sometimes incorporated into theories like string theory, as discussed later, that mathematically try to set up other hyper dimensions beyond our own physical universal dimensions of space and time. For instance, some models about consciousness deal with other hyper dimensions outside of space and time in almost the same way; this would imply that our minds have access to other dimensions in some way. And this idea would open up all kinds of possibilities about how we could have access to dimensions that include the realm of God (through consciousness for instance). And then, questions of eternity, where did we come from, where are we going, how did we get here, and so forth would actually go away, as long as we have incorporated into this idea that God belongs to a realm that is independent but yet a superset of our existing universe of space and time.

So that brings us back to the concept of time. Since omnipotence implies ultimate power, an Omnipotent Creator such as God would imply that God can do as God pleases to do. So then, in this case, it should be no surprise that God can create space, and God can create time. He can create positive time universes; he can create negative time universes. He can create mass, he can create anti-mass. He can create water, he can create fire, he can create wind, and he can create earth. (That's my tribute to Empedocles and Plato.) And God certainly made that clear enough in the burning bush when he had Moses shaking in his boots (except that God made Moses take his shoes off) on the mountain as he told him "I am that I am". It could be supposed that God didn't want to go into details at that point since there were more pressing matters at hand, for instance what to do with all those rebelling Israelites creating mayhem below. But he probably needed to give Moses enough of an idea about just how omnipotent he was anyhow. And it probably worked. The result of course, according to legend, is the first five books in the Old Testament of the Bible and its Ten Commandments, written by Moses and serving as God's desire for the people's obedience to him. Muslims, Jews, and Christians can all share agreement on that. What they might not know, however, is that I had a dog named Moses once; and he wasn't that obedient!

In the biblical context of time, Genesis starts off well enough with those profound words "In the beginning God created the heavens and the earth." Not a bad way to make a point soon enough. We've discussed the relative creation already, and will come back to it again. Only problem is that the Bible doesn't say much about how it all ends. We have to wait for the biblical New Testament writers and the disciples to ask Jesus about that, except for a little nibble from Daniel in the Old Testament. And not surprisingly, they all get pretty coy about that. Jesus tells us only to start worrying when we see people heading for the hills. But they all like to talk about a time when that "desolating sacrilege" is not where it's supposed to be, in which case you should really start to be worried.   Well, unfortunately we can only speculate what he meant by that desolating sacrilege; and we'll have to leave that one to the prevue of exotic theological theory.   But Jesus does throw out one more useful tidbit.   He says the sun will be "darkened"; and there will be much "tribulation".   Well, on this point we can all certainly agree; if we lose the sun, we earthlings are certain to face tribulation.   So if we look up our reference book on the sun, we find that the sun has about 5 billion years of fuel energy left to continue radiating the earth with its continuously reabsorbed gamma rays resulting from the fusion of hydrogen into helium under the sun's surface.   Now, 5 billion years is about twice the amount of time that it will take the Andromeda galaxy to come careening into all of us here in the Milky Way galaxy, which will make for a fine mess, if that's any consolation.   So then what?   Well, besides probably being a little longer than our lifetimes, we don't have to worry much. Like the earth, the sun is approximately halfway into its expected lifespan of 10 billion years; so it's still got a long time to go.

Another possibility that Jesus was warning the people about was that the sun, near the end of its lifetime, will, for all intense purposes, become a red giant; that's the name given a star that loses all its hydrogen fuel in its core so that the outer layers collapse under gravity, thereby heating up the core to produce a much larger outer shell.   It would get much redder (hence the name), if in fact that's what Jesus meant by getting "darker".   In any case, this eventuality would make things a little hot on earth; and in fact the earth would be nearby enough to burn away during the mayhem.   At least that would work for the biblical apocalypse enthusiasts.   Even St. Peter in his second letter of the New Testament warns us about the earth's eventual destruction by fire.   And it turns out that the Bible warns us many times that the earth will end by fire; so that much would be consistent with the red giant hypothesis.

However, apostles of Jesus at the time thought he was talking about the end of the world, and Jesus' second coming, as happening in their lifetimes. But it didn't happen that way. Jesus was of course talking about a time further down the road, much further in fact; but he probably wasn't talking about 5 billion years down the road either.

Humans would be quite likely to have destroyed themselves by then anyhow. Perhaps one could get a better feeling for the time of our demise by just extrapolating how long it will take us to use up all our natural resources, or the point of no return when pollution damage to the earth is so bad that greenhouse gases absorb the planet into a cloud of smoke. Just look at what has happened to our endangered animals and precious environment. Things could get pretty messy for us too before they get better, or worse depending on your vantage point.

It's too bad how man has treated his world in the face of our rather seeming special status as a stable orbiting planet in the universe; we'll see just how special that status is later on. When people of faith ask why God allows bad things to happen, you can well guess that the message being conveyed is of much greater importance than that which has been lost. The use and destruction of our natural resources, along with continued global warming and destruction of our climate, due to our increasing numbers of chemical pollutants in our air and water, causing the resultant loss of sunlight and ocean habitats, could also have been what Jesus was talking about. But what we do know for sure is that there is an upper bound on our planet's life sustaining ability not lasting more than another 4 billion years; and human intervention will push this bound lower, not up.

As for time in the cosmos, it's expanding apart itself at an accelerating rate; in about 10 billion or so years, it too may have problems supporting life as we know it anywhere. The cosmos' expected "lifetime" really depends on how you interpret that cosmological constant now used by some cosmologists to explain the recent surprising accelerating expansion of the universe. Einstein thought the constant was his biggest blunder, you recall. But some scientists today see it as a neat mathematical trick to explain all that dark energy, which no one has seen, but which would cleanly explain the universe's seeming repulsive gravitational field that gives rise to the expansion in the first place. The problem is, depending on the model du jour, you get very different numbers for this constant; and none of these numbers are really understood to explain all that dark energy and repulsive gravity that only seem to exist in the outer cosmos. Well, in the end, there really are only three scenarios that are possible. One is that the universe will finally slow its expansion, stop, and turn around, leading to a Big Crunch, where we all eventually become one big happy family of pancakes destined for the center of a universal black hole. Two is that the universe expansion slows down, and stops (or at least remains stable without acceleration in either direction), leading to a Big Stall. And finally, there is the ever expanding universe that seems to continuously accelerate apart, until eventually, all atoms in the universe break up and there are no more stars left to keep us warm, leading to a Big Chill, not that there would be much of us left together to keep warm. As we have seen, the data from type Ia distant supernovae in 1998 and the WMAP probe of the cosmic microwave background in 2003 support the latter of the three scenarios. After the universe expands apart, it is left to the prevue of philosophers to discuss whether the cosmos would still exist; but in this case it would have been transformed into a phase of no return, as remaining space dust.

As the fabric of space expands and cools down, one can philosophize how time would be affected, guided by the laws of relativity. You see, when an observer sits at a distance watching a massive black hole swallow everything in sight, he would notice that time for the poor subjects falling into it would be grinding to a stop. However, as the universe continues to accelerate, it behaves like a black hole in reverse; time in the universe would be seen as speeding along to an outside observer, as it would be *his* clock that would be stopping. The outside observer such as God would see virtually no clock movement at all. No wonder God set things up like this; it seems he doesn't age. And I seriously doubt that he would even need a Movado for Christmas, thank you very much. Also he would have no need for an alarm clock; God it seems has "all the time in the world" to wake up in the morning, whatever morning was!

Now there are serious consequences to this geometry of the universe, as predicted by General Relativity. The universe horizon, as it expands and accelerates, becomes a snapshot of the big bang, namely it remains at time $t = 0$. The outside observer doesn't age, and actually could be going backward in time, just depending on what antigravity field, if any, he may be sitting in. But a person inside the universe trying to reach the horizon would have to be exposed to an antigravity field in order to get there, which you'll recall caused the universal expansion in the first place, according to inflation theory. Well we can see how that might make sense; after all, how would you suppose to get to the beginning of time from where you are now without somehow feeling like you are going backward in time. The only way to do that is to have access to an antigravity field; we'll discuss this possibility more when we talk more about the very real physical possibility of the existence of wormholes, and what that means for intergalactic travel or chronology. Another outcome of this idea is what might have happened to the speed of light since creation; and it might not have happened the way you had expected, especially if you're a physicist! Again we'll be getting to that later.

Seriously, one can speculate what it means theologically to have an "open" universe not subject to collapse. Of course, one could speculate what it means theologically to have a "closed" universe, also, that does collapse! For one thing, it does seem to suggest God plays for keeps; namely, there's no starting over. Once you have a record, you always have a record, and that sort of thing. This idea fits well with religious conservatives; and indeed an "open" universe works well with Dante's Inferno, where the inner circle becomes a frozen wasteland. But the idea of an "open" universe might make some scientists who are searching for that ultimate symmetry in the universe a little uneasy. For instance, how do you recover what has been lost? The concept of symmetry violation in general, as we'll see with the CPT conundrum, won't make them feel any better. For some scientists, the idea of a cosmic yo-yo, where the universe is constantly destroying and then reinventing itself back into existence every so often, is more appealing, a kind of periodic reoccurring big bang scenario as it were.

There is a speculative model put forward by Stephen Hawking, which he calls the "no-boundary" approach. In it, he suggests the universe may not have had an instant of scientifically unexplainable infinite energy-space-time singularity that gave rise to the big bang after all. He suggests that time may be alternating between positive time and imaginary (negative) time where the cosmos at each time switching point physically expands from a small point, stabilizes, collapses back to a small point, then switches time direction again, and so forth. Visually, a sphere would represent the space of the universe while time is circulating around its circumference with the north and south poles acting as time switching points. Hawking then continues that if this model were correct, there would be no need to ask God about how he created things, because the cycle would be, and would have been, continuous throughout all eternity, forward and backward. Unfortunately, this idea doesn't really explain where space and time came from, and how this endless motion came to be oscillating in the first place.

But God as Omnipotent Creator wouldn't need to worry in either case. If he exists in an outer realm that doesn't depend on the energy and gravity of our universe (but which could include it), then he wouldn't be dependent on our universal space and time either. And he would maintain his eternal presence over the deep or void, as the first chapter of Genesis suggests. As inflationary theory, no-boundary theory, and the postulates in previous chapters suggest, there really is nothing to stop the whole universal process from restarting from scratch anyway, although these theories model the process starting from within the context of our present universe while the postulates here suggest the process may be controlled from the hyper dimensions of an outer realm. Now all of these ideas could support parallel (or even negative) universes; but I'm guessing that Giordano Bruno, or theologians in the Roman Catholic church for that matter, wouldn't want to hear about that anymore.

But what they might all be willing to hear about is the Anthropic Principle. In its weak form, it states that all possible states of the universe are not equally probable; rather, there is a state of the universe that must exist which meets the absolute requirement for sustained life to exist in it, to evolve in it, and to be observers of it. For some cosmologists, however, this doesn't go far enough; so they have developed a stronger form of the principle. The Strong Anthropic Principle says that intelligent life must come into existence in the universe; and once it is there, it will always be there and do what is necessary to stay there. That is, life will always find a way to hang around, develop, mutate, and evolve. We will see what opportunities there might be for an advanced civilization to exist in a dying universe when we talk about the possibility of the existence or creation of wormholes. But we'll also see that wormholes aren't exactly what they are all cracked up to be; and they introduce all sorts of possible chronology difficulties that have puzzled people from Einstein to Hawking.

These concepts can get kind of heavy, as Marty McFly would say in ***Back to the Future***; but there are serious implications for the universe as a whole when applying the Anthropic Principle. For instance, as life matures and gains command of its universe, which thus naturally maintains a state that is conducive for its existence and growth, life will eventually consume and take command of all the forces and energy that the universe has to offer, and spread out to all possible regions of its expanse, and indeed into all the possible parallel universes in existence. That would imply of course that the universe has to be "closed" in order for life to reach everywhere, which all available evidence now seems to show that the universe isn't "closed". But we'll see later that there might be a possible solution to both the Pioneer anomaly and dark matter issues in cosmology that imply there is a dimension in which the universe does become "closed"; even string theory gives some indication of this possibility in the way it models hyper dimensions of the universe that wrap around themselves. Meanwhile, we're stuck with what we can see in the visible universe, which continues to show an accelerated expansion and flatness in the cosmos. Of course, what is visible versus what is not is up for debate; there could be a lot out there beyond what we think. The flatness phenomenon and inflation theory certainly give some indication of this. As for all those other universes, the Anthropic principle may allow for them; but then Giordano Bruno's observation is not only still valid, but even more so. God has enough trouble making us behave in this universe; consider what he would have to do with multiple universes if they exist. And we still are asking ourselves whether there are life forms on other planets within our own universe, much less other universes; but we're getting to that.

As for what we do know that exists, namely our universe, there is something God did with it to make it self sustaining, perhaps even a harder job than it was to create it! A fascinating tool was used; physicists today call it quantum mechanics. It would be a subject that Einstein despised and nobody else really understood; but it was a subject that we only started to learn about during the past century. Prominent physicists today still confess that they don't know exactly why it works. But the most credit for its discovery would really end up going to a man who became a leader in Nazi Germany's atomic bomb project, no less; but before you pass judgment, it would be a project which he said after the war that he tried to sabotage! To be sure, fate must back him up because the discoverer of God's favorite tool for forming the universe could have never remained onside with the Nazis; it is the tragedy of one man to love his country, but hate its regime. (That he was once denounced as a "White Jew" by some Nazis for teaching Einstein's theories, and that he seemed to learn rather quickly after the war how the Americans built their bomb while still being interned as an Axis scientist, might add validity to the case.) Yet, his discovery would not only be the favorite tool of God the Creator, it has become a favorite tool of modern man, who now uses it in over 25 % of the world's GDP output. Quantum mechanics is the next clue that we discuss in trying to understand the revelation of God within the perspective of our universe.

VIII. Playing with Dice

Classical physics as we knew it in 1900 had much success in describing macroscopic events, electromagnetism, Newtonian mechanics, thermodynamics, and gravitation; but it had a basic flaw; it could not describe the function of an atom. Later in the 20th century, it also had no explanation or mechanism for the seeming uniformity and flatness of the entire universe, albeit it could describe orbital planetary motion, thanks to Newton. In classical physics, the electron orbiting around the nucleus of an atom should really simply spiral out of control toward the nucleus and be absorbed; but in the real world, that's exactly what does not happen. And a good thing too, since most things in our world depend on electrical interactions of atoms in order to hold themselves together in one way or another, including the atoms in our own bodies. Even with the advent of relativity theory, physicists could not account properly for expansion of the initial minute universe; it should have just caved in on itself right from the start, thus the impetus for Einstein introducing that cosmological constant that seemingly came out of nowhere which cosmologists are still trying to use today to describe the acceleration of the expanding outer universe. So what was missing?

Einstein spent a lot of time grappling with issues surrounding the creation of the universe in his relativity theory. In his own quest to the find the Theory of Everything, he was troubled by the evidence, and indeed by his own equations, which seemed to show that the universe began from a point of singularity. He also didn't like the idea that the universe was seemingly driven by random means. That's when the subject of God would come up in his writings; but he resorted to it rhetorically when he would make a defense of his own beliefs. "God does not play dice with the universe" was one of his most famous lines. Of course Stephen Hawking would say later that maybe God does play dice after all. Knowing something about the humor of God, I would perhaps agree with Hawking on that score at first glance; but we'll see later who would have probably won a hypothetical bet between the two. Yet there might just be some fundamental, and maybe even a humanly understandable, if not humanely explainable, force behind this universe; and dang it, Einstein wasn't going to find it! Together with theorists like Schrodinger, Heisenberg, Bose, Dirac, von Neumann, and others, Einstein helped to shape, if unwillingly, a theory that has helped to explain most of the remaining unresolved issues on all things microscopically universal; it came to be called quantum mechanics. The theory would forever haunt Einstein because he never really accepted the idea that physics could operate on chance, which quantum mechanics naively seems to do. Because of its important role in the creation of the universe, you will excuse me if I take you on a short recourse of modern physics.

In 1905 Einstein published a paper (while polishing off his Theory of Special Relativity) on the photoelectric effect, where light was absorbed by materials that were in turn releasing electrons at discreet energies. He hypothesized that light itself had quantized energy proportional to its wave frequency; for this work, he would end up receiving the 1921 Nobel Prize in physics. Separate experiments were being done that were showing light behaving as wave or as a particle, depending on the experiment. Physicists wondered how this could be. Further, a student named Louis de Broglie (1892-1987) published a thesis that extended a theory of light particle/wave duality to all particles, including electrons. So then along came one Erwin Schrodinger (1887-1961) in 1924 with a paper that mathematically described the hydrogen atom using operators and wave functions, thus creating his infamous Schrodinger's Equation. At the same time Satyendra Nath Bose (1894-1974) published a paper with help from Einstein that proposed different states for the photon where there was no conservation in the number of photons. What was finally understood was that light sometimes appears to act like a wave and sometimes to act like a particle because of the accumulation of light particles distributed descriptively by a probability function that predicted where each particle should be and sometimes having the same energy state as other light particles. Then Werner Heisenberg came along in 1925 with a postulate that became the fundamental doctrine of quantum mechanics: the process of measuring the position of a particle causes a disturbance in the particle's momentum, due to the photon that bounces off the particle and needed for the measurement to be made, such that the change in momentum times the change in position must always be greater than or equal to some very small constant number, called the Planck constant by physicists and named after German physicist Max Planck (see below). The theory became known as the infamous Heisenberg Uncertainty Principle; and the person who was last to accept it was none other than Albert Einstein. He finally agreed to accept it however when he was convinced that the error in measuring a position translates accordingly to the error in measuring time due to the constant velocity of light used in making the measurement. The Heisenberg principle arises from the boundary conditions set forth by the particle and wave duality of nature. With some formulism added later by John von Neumann (1903-1957), the theory of quantum mechanics, and the ability to describe all particles as waves, including photons in light and electrons in atoms, was born. For his accomplishments in the field, Werner Heisenberg was awarded the 1932 Nobel Prize in physics. Later, as a leader in Nazi Germany's atomic bomb project, he said he saw to it that the project never really got off the ground; and he wasn't talking about position versus momentum! His epitaph reads "He lies somewhere here."

One of the interesting consequences of quantum mechanics is the nonzero probability for particles to tunnel through barriers, due to the wavelike description of the probability for a particle to be at any place at any particular time. In terms of macroscopic absurdity, one could thus stand next to a wall thinking that there will eventually come a time when he or she would magically appear on the other side, like Superman I suppose. In any case, quantum mechanics actually says that there is some nonzero probability for it to happen. But you should be forewarned; the probability for even an atom in your body to make it to the other side is quite low, much less for all of you. Thus the probability for all of you to suddenly reappear on the other side of the wall is a multiplication of the probabilities for all particles in you body to do so, much less the need for all those atoms to turn up at just the right spot in order to maintain your body's congruence; otherwise, you might end up as some kind of zombie or be split between both sides all at once! And either way, that number is likely to be quite low! Yet you always have some clown in physics class who stands there by the wall to make the point! Later we will speculate on why this seems so strange to us inside our space and time, but actually may make a lot of sense when the entire set of possible dimensions of the universe is factored into the equation.

Another consequence of quantum mechanics is that it assigns a discreet probability for all possible events that could occur, with the most likely events to occur the most often, and so forth. The impact of this theory on 20th century technology and onward has been nothing short of phenomenal. With the advent of semiconductors and microchips that really work on this go-through-the-wall-like principle, called quantum mechanical tunneling, over 25% of the world's present day GDP is based on quantum mechanics, from the processor that runs your computer to the alarm clock that wakes you up in the morning to the toaster that toasts your bread in the morning to the ignition switch that starts your car in the morning and so on. It works because of the sheer numbers of particles involved in making the possible hop across boundaries; there will always be one particle that may succeed.

Now as this principle of quantum mechanics applies to everything in nature; it also to biology (after all our bodies are made up of atoms with nuclei and electrons like anything else). An interesting facet of this theory is that it can be blamed at the cellular level for causing human diseases like cancer and mental disorders. It could also be used to explain hormonal impulses that vary from person to person and why glands behave differently for each person. For instance, one to three percent of American males in surveys admit to being homosexual. Strictly from the viewpoint of quantum mechanics, this would not be a surprise. It might interest those who wish to legislate against this human behavior that the whole process which leads to sexual orientation could easily be explained by natural laws of quantum mechanics acting through the biology of our bodies. For instance, we know that everyone has a different metabolism that makes some people fat and some people thin; yet we don't stop them from marrying each other! There will always be a small fraction of cellular activity in our bodies that behaves differently from the rest; so that when we show prejudice against those that have a natural behavior different from our own, the tables of joke and judgment then begin to turn on us. The fact that God wrote quantum mechanics implicitly into the biological, chemical, and physical code of the universe (and our bodies) shows that he had a will, purpose, and desire to create some random diversity in the universe, while at the same time he designed it in such a way that more favored events would occur with increased probability, the more the desired, the higher the probability that they would occur. We can see this in the diversity of people's faces, in the diversity of scenery and places, in the diversity of everyone's unique personalities, and indeed in the diversity of the stars and the universe itself. We can philosophize forever why God did this; but one spin on it would be that he wanted a mechanism for people, for instance, to have to deal with a variety of events and issues, some with even low frequency or occurrence, in order to see how we could (or could not) handle them, while allowing the more desired events to occur more of the time, such as giving people the hormonal instincts to recreate and propagate the species. In the stars, we see unusual events occur also when certain stars become supernova or when black holes form, giving rise to all sorts of intergalactic mayhem and beauty; yet most of the time, most of the universe remains quite tranquil.

The fact that the more desired events occur with more frequency seems like something out of Charles Darwin's evolutionary theory of nature; except that he wasn't using quantum mechanics to arrive at his treatise! That's too bad, because he missed an important point in his theory. Evolutionary theory really ends up being a subset of quantum mechanics, such that evolution, as it applies to biology and genealogy and so forth, should only predict what quantum mechanics would predict. Unfortunately for Darwin, quantum mechanics will not predict necessarily the survival of the fittest or the cosmic regeneration of only the best elements of a species; rather it simply predicts the most likely survival of the most probable set of events and regeneration of the most desired phenomena of nature. In order to find what is most desired, we only need to count what there is in most abundance in space and time; and that distribution of nature which is in abundance versus what isn't has been coded into the fabric of the universe from the day of creation, according to the laws of physics and the resulting symmetry breakdown of the universe which occurred from the initial Planck interval after the big bang to the present. Hence the symmetry breakdown that has occurred in our finite universe causes evolutionary theory to ultimately fail in its prediction of the eventual rise of more and more perfect species. Darwin confused the most desired features of the universe with its strongest features; and they are not necessarily one in the same! And that's why, for instance, Darwin in his later years scratched his head trying to understand why an advanced cosmic aged species would only end up fighting for its own survival against the coming doom of the cosmos. He simply hadn't applied the rules of quantum mechanics to the game; in fairness to Darwin however, he would have had to wait for another 70 years after his treatise for quantum mechanics to come on the scene! A quick example of this problem can be explained with the human brain. Scientists have detected that the human brain has increased in size over thousands of years, giving rise to smarter and smarter people. As the brain continues to increase in size, the more likely that over time humans will develop problems trying to reproduce, even to the point of dying out. Bigger brains may make us stronger, but quantum mechanics would then suggest that they are not necessarily more desired.

Another example of the rules of quantum mechanics being applied to the most desired in nature is the issue of what has happened to all the stable planetary orbits in the universe. So far we only see a few stable planetary orbits in all of space; and they are all here in our solar system! Theologically, we can speculate why the earth has the unique stable orbit that it has, as the geocentrists of the 16th century and on backward did. Investigations of newly observed planets in our galaxy and the nearby Andromeda galaxy show that the probability to have life sustaining stable orbit planets about a star seems quite low, as we'll see later; but as the earth shows, it is at least nonzero. Obviously, one can conclude that God didn't want to have many planets like earth around, if only perhaps to be faced with the predicament of Giordano Bruno! But one can certainly understand why he wouldn't want too many black holes around either. And certainly one would understand why he would want so many beautiful and majestic stars in the heavens, the more, the merrier, in fact. On this last point, the geocentrists might have a problem however; it seems they also needed space for heaven and hell, while the earth had to fit in the center of things, regardless what else was out there!

Well the main reason this discussion of quantum mechanics becomes so important is to help us understand how God used it to create everything in the first place. If we recall back to the theory of inflation, the whole universe, so goes the theory, was created completely out of a vacuum fluctuation of zero ground state energy. This is basically a nasty way of saying that God created something out of nothing. Indeed, if we are to believe Einstein's General Theory of Relativity (and why he suddenly stopped in the middle of a busy road one day to think about it), the entire positive energy due to the mass of our present universe is negated by the entire negative energy due to its gravity (negative because gravity is an attractive force). According to the law of conservation of energy then, all the energy in the present universe is consistent with that which existed before the big bang; namely 0 should always be 0! Therefore God didn't break any law, especially since it was his own not to break! What he did do, however, it seems, was to make use of his pet tool – quantum mechanics. The vacuum fluctuation of nothingness, wasn't exactly nothingness after all; there would have been a sea of virtual particles and antiparticles that could cancel each other out (thus preserving the conservation of energy); but a quantum mechanical process known as tunneling may have occurred in the early universe, as it is proposed by inflation theory proponents. By early, we're talking 10^{-35} second from t=0. In this idea of the universe's creation, a quantum mechanical energy fluctuation occurred during what amounted to an imaginary time interval that caused these virtual particles/antiparticles to have a real state of *nonzero* energy; the imaginary time interval was important in order for God to observe his own law of energy conservation. In this interval, gravity of these early bird particles was actually repulsive such that they split apart in what amounted to 10^{25} meters of space from less than 1 cm. The virtual nature of these particles allowed them to be faster than light; but as they became distant, a phase change in the early universe occurred that turned them into real particles (the kind that make up you and me) which can only travel at most at the speed of light, so goes the theory.

The reason the whole process is quantum mechanical is that Heisenberg's Uncertainty Principle is called into play. Recall that it states that the measurement of an object's momentum change and position change could not both be known with a better resolution than the value of the Planck constant (10^{-34} joule-seconds, if you really want to know). Another way to put this (and what made Einstein finally give way to the theory) is that by measuring an object's energy fluctuation (instead of momentum change) in some unit of time (instead of position change) and then multiplying them together, you must get a value greater than or equal to the Planck constant. Now we can have some fun. Given the time interval of the universe (13.7 billion years), the energy fluctuation needed at creation to make the universe come out as it did would have to be greater than the Planck constant divided by 13.7 billion years, which is 10^{-41} times the equivalent mass energy of a single proton! It actually gets less than that over time, since the universe is not assumed to be coming to an end just yet. As I said, God has a pretty useful high resolution tool with quantum mechanics! And that something is pretty wickedly close to nothing!

By the way, the tunneling terminology occurs because of the quantum mechanical nature that allows everything to be described in terms of a nonzero probability for existence in some space and time. If there is a seemingly insurmountable barrier in the path of a small object rolling toward it, the theory describes a nonzero probability that actually exists for the object to simply hop over it and be on the other side at some point in time. The Planck constant itself is named for a German physicist named Max Planck (1858-1947) who introduced it in 1900 for his accurate quantum formulation of the distribution of the light radiation emitted by a blackbody, for which he received the physics Nobel Prize in 1918. As for a blackbody, it is anything that is a perfect absorber of radiant energy, and something, I might add, that I had to make for an undergraduate research project. Now, if you are really interested yourself in understanding what a blackbody is, try closing the windows in your bedroom some hot summer morning when the sun is shining in and the air conditioner is turned off; and you'll find out quite quickly what a blackbody is!

Now I wish I could tell you that everyone is happy with this theory of creation, and all things are resolved; but the usual friction that sometimes still exists between science and religion is always there. Some will say that when religion answers the question, then science seeks to question the answer; but it's also true that when science provides the answer, religion seeks a new question. For instance, where did this quantum epoch of virtual particles come from anyway, why did they suddenly inflate, and what's the story with the vacuum they were sitting in anyhow? Physicists are great at finding a better understanding about *how* the universe went from point A to point B; but they are not usually as swift to understand *why* the universe went from point A to point B. Stephen Hawking has proposed the idea that the universe might be completely self-contained with no absolute boundaries at all; and therefore there would be no such thing as creation or destruction. And Einstein was troubled with the notion that creation might be divinely inspired. Thus, like most physicists, both Einstein and Hawking don't seem to like the idea of appealing to God for the initial conditions of the universe, because at the very least they would prefer to use the laws of physics to get there. On the other hand, theologians haven't been much help; for instance, even Pope John Paul II did not want scientists to study how creation started itself, because that belonged to the "realm of God". So while theologians don't seem to care about the how, because they seem to know the why, scientists seem to almost dismiss the why, as long as they know the how.

Well, sometimes good people just simply disagree, not only about the how versus the why, but also about each individually. Some theorists come up with theories such as string theory, the idea that everything comes from a 10 or 26 dimensional space. But some accomplished physicists disagree with that approach also. They go on to admit that even quantum mechanics itself can't be used in its present form to explain the entire inflationary process. Something is still wrong with the math, whether it's in the inflation model or quantum theory or somewhere in between. Where there is widespread agreement among many physicists is that there must still be some new physics yet to connect the dots from relativity to cosmology to quantum mechanics. That's another way of saying physicists haven't yet put the whole puzzle together. And thus a postulate here is that physics will eventually reach an endpoint boundary when after all the new physics and all the new discoveries are incorporated, there will still be something beyond the physical that is needed in order to reach a final state of completeness. I'm sure that many physicists will not want to hear that; but to simply ignore or dismiss this possibility is evidence of bias, and thus risk never getting to a final Theory of Everything. We will see later, when we talk about how mass comes about, that there is one perfect example of this fallacy, the resistance of physicists to embrace the theories of one absolutely brilliant but recluse German theoretical physicist who may have actually come close to a Theory of Everything before he died, but was shunned due to an unwillingness to publish and a disability that kept him out of academia for most of his life. And we're not talking about Albert Einstein. But we'll see however that it has been many of his ideas for constructing a unification of all forces of nature that are now being used in the developing theories of quantum gravity, such as string theory. His ideas have also started new ground into the idea of incorporating spirit and consciousness into models of a multidimensional universe that includes, but is also beyond, space and time.

There are other examples also where we can find resistance by scientists to the idea of incorporating theology with theories of creation. For instance, in the first edition of Stephen Hawking's *A Brief History of Time,* Carl Sagan wrote an introduction in which he concluded "This is a book about God...or perhaps about the absence of God". Now, it turns out that Hawking never said any such thing; in revised editions of the book, Sagan's introduction has gone missing! The problem here of course isn't that Hawking would allow Sagan to write the introduction; goodness knows what he was going to write. The problem is that Sagan seemed to show contempt for God's place in the universe and then put those words in Hawking's book! The closest we actually find Hawking talking about the "absence of God" in his book is when he talks about one theory of creation that has "what place, then, for a Creator"; and even then, Hawking appears to be taken out of context. Of course, in some of Sagan's own writings, we also find what some would consider contempt for religion in other places as well. The point here is that Carl Sagan was indeed an accomplished and famous scientist, Cornell professor no less, and well regarded in the physics community. And he could easily be seen as someone leading a larger class of scientists to see religion as an obstacle to science, and not as a partner. It is true that in Galilean times, the church certainly earned its credentials of obstinence; and it certainly takes time to heal all things. Yet one can't help but observe that scientists make a big mistake by ignoring religion's abstractness and majesty, just as religion has continued to make the very real mistake of ignoring scientific observations and facts when they just don't fit the doctrine, such as the creation timeline, geocentricity of the earth, biological patterns of people, evolution, and so on.

The real culprit here with the scientific community, like the theological community before it, is intellectual arrogance. If pride is the root of all evil, then arrogance runs a close second. When arrogance enters into religion, it is called blind; but when arrogance enters into science, it leads to prejudicial bias. And as all scientists know, bias corrupts the result of any measurement of scientific data because it causes the scientist to pull the result in a direction that he or she wants it to go; but the data itself may or may not want to go there. And thus the scientist can easily lose his or her reputation of truth and purity that he or she trains so hard in graduate school to achieve when such mistakes are made.

As for the creation story, it is clear that new physics will be needed in order to fully align relativity with inflation with quantum mechanics; it could be well guided theology, and may in fact have to be at some point. There are still inconsistencies in the way quantum mechanics explains things versus the way relativity explains things versus the way general cosmology (e.g. inflationary theory) explains things. Quantum mechanics and General Relativity will undoubtedly continue to play leading roles in the creation story, as they have continued to be successful in describing microscopic and macroscopic rules of nature, respectively, and how all those minute particles and giant objects in the cosmos manage to get along and hold our universe together.

One way these theories could find common ground and incorporate theology is through the introduction of other dimensions beyond space and time, as mentioned already. For instance, if quantum mechanics really is describing a projection onto space and time of a higher dimensional universe, one might see these projections in space and time as nothing more than probability distributions for something to be at point (x,y,z) and some time t. The higher its probability to be seen at these space and time coordinates, the more time it might be actually spending along another dimension while at these same space and time coordinates of our universe. Similarly, General Relativity describes the universe in four dimensional space and time; if one extended the theory out to other dimensions, one might get a very different idea about the theoretical construct of the fabric of a multi dimensional universe beyond, but including, space and time. Hence, with the inclusion of extra dimension(s), both theories might find exact solutions for particle interactions, energies, and expectation values and find agreement with each other. And this certainly would have made Einstein very happy, because, not only would he have won the hypothetical bet with Hawking, it would mean that God doesn't play dice with the universe after all; particles interact and exist in exact states that could be predicted and verified. The seeds of these ideas are planted in string theory that is still undergoing development today in order to find the ultimate Theory of Everything.

And even after all the new physics out there is discovered and new theories are verified, it will no doubt be just a matter of time when physicists will indeed put the pieces of the puzzle together; and they will fit perfectly except that somebody will notice that there is still some last piece missing. And the reason it will be missing is that it will likely represent an interface to that which we can't see in space and time, and thus never be actually measured or discovered, just theorized. And that is where theologians could be helpful to step up to the plate. Let's hope they will get on with it and allow the scientists to show them what they got, and vice versa. Then the real fireworks will fly, because it may become necessary that a complete Theory of Everything will need to incorporate ideas and issues from across disciplines. One of the reasons it doesn't happen of course is that cross territorial boundaries create natural friction between various professions. For instance, you don't here about many physicist conventions with psychologists, or many psychologist conventions with cosmologists, if, for no other reason, they fear the result will take on some new form of astrology or psycho babble! But we will see that serious theories under development by theoretical physicists and psychologists are leading into areas where perhaps they did not intend them to go, and may still prefer that they did not go there; yet a complete Theory of Everything may end up requiring them.

Valentinian mythology supposed that the breakup of the original perfect nothingness of the heavens led to the finiteness of our present universe; in other words, we are separated from God by a fall from grace, by being here in our universe. If this is the case, then it could be argued that the primordial tunneling of those virtual particles from the "perfect" vacuum represents the arrival of sin, because they gave way to a very imperfect and asymmetrical universe. But it doesn't stop there, because even if those particles tunneled through to a real nonzero vacuum state, they would have still paired up with their corresponding antiparticles and annihilated each other! And that almost happened, but not quite. By a rate of one per billion, we have the existence of matter over antimatter; the rest did finally annihilate. And voila, our universe came into being with a clear bias of matter over antimatter. It took physicists a long time to figure that one out; and they're still figuring, as we will see. What could be the theological interpretation of this symmetry breaking, what allowed this to physically happen, and how could we ever get, or even want, the symmetry back? The physical answer lies with something called CP and CPT conservation, discussed here after we talk about mass, and the conundrum that caused its violation. Thus, the only way that the universe can ever be fully reconciled back to God, implied by Valentinus, is in the restoration of universal perfect symmetry; and the regaining of CP and CPT conservation would be required. There are serious efforts by physicists today to find this symmetry, and what extra dimensionality besides space and time it might need, although their motivations are not so much a search for God as they are a search for the Theory of Everything. But keep in mind that the search for God is itself a bad news good news scenario. With the restoration of perfect symmetry that could lead to God, as Valentinus would require, our present physical universe could not exist; so one has to be careful what one hopes for! As for those physicists working so hard at it, they would be probably more satisfied just to understand it than to wish for it. On that score, we can't blame them; but the side benefit for them is that maybe quantum mechanics would one day be turned into a perfect science after all, albeit we wouldn't be able to see it from within the confines of our own physical dimensions of space and time. But it might have made both Heisenberg and Einstein simultaneously happy, although maybe not personally with each other.

Perhaps there is yet another angle from which to look at the problem of the asymmetry of creation; one that could even stir the interest of theologians. It comes from particle physics; and it's a theory based on the standard model of physics (e.g. the theory that has most acceptance) that tries to explain how mass itself comes about. We'll see however that mass may not be all that it seems to be; but particle physicists have a working theory for it nonetheless. And there is some fairly expensive equipment, enough to put world governments into the red, being built to confirm it. If we understand mass, we might just understand gravity a little better; and if we understand those two, we might just come a little closer to a fuller Theory of Everything, and maybe of the asymmetry of creation also. And just to perk the interests of the theologians, it seems, is the particle that some physicists are calling God, because it gives rise to what we are. That discussion is next.

IX. "The God Particle"

In 1988, Leon Lederman, the Director of the Fermi National Accelerator Laboratory (Fermilab) in Batavia, Illinois, shared the Nobel Prize for physics with Jack Steinberger and Melvin Schwartz for the discovery of the muon neutrino and its feature to have a doublet pair particle. He was the Director while I was finishing my research there for my Ph.D. that I received in 1989; and I am forever indebted to the laboratory for giving our collaboration the final data run that we needed which allowed me to finish my thesis. I often joked that Leon beat me by 10 months. Unfortunately I never worked with Leon, except once being asked by my collaboration's spokesman to spontaneously convert some decimal numbers to octal format for him over the phone. During the next several years, Lederman was a great proponent of the Super Conducting Supercollider (SSC), being built in Texas with its own directorate, separate from Fermilab; only problem was that the U.S. Congress had other ideas. In their infinite wisdom, Congress cut off funding for the SSC lab a few years later in 1993; perhaps, they weren't supposed to have a Christmas party. Back at Fermilab, Lederman continued to express the benefits of the SSC project to the public; and in the course of so doing, he wrote a book about a particle that physicists hoped to find there. The book was called *The God Particle.* In it, he uses his fabulous humor to describe the life of a physicist, including his own career, while describing the importance of finding this particle, which was the subject of the book.

At the beginning of his book Lederman starts out by admitting "We don't know anything about the universe until it reaches the mature age of a billion of a trillionth of a second. That is, some very short time after creation in the big bang. When you read or hear anything about the birth of the universe, someone is making it up-- we are in the realm of philosophy. Only God knows what happened at the very beginning."   In this period after creation, the universe existed as a sea of very hot quarks, electrons, and their antiparticles which were free to roam around and unable to bind together because of the heat; but particles such as these that had rest energies down to the order of 1 MeV (mega electron volts) could theoretically exist, according to Heisenberg's Uncertainty Principle. One of the ways we can find what existed at this era of time is by observing collisions of beams of particles going near the speed of light at giant accelerator laboratories, such as exists today at Fermilab near Chicago, and the Center for European Nuclear Research (CERN) in Geneva, Switzerland.   As we try to nail down history before this period, we are in the realm of theory, philosophy, and theology, because it is very hard to prove by experiment what actually occurred. The best we can do is to extrapolate scientific thought based on the scientific data that we do have to that period in time that we don't have. There is often no scientific data because the particles that we try to observe today through particle physics experiments, not only just don't hang around normally in nature, they only existed during the primordial epoch of creation itself.

As for creation, the same problem (from a physicist's point of view) unfortunately occurs because something unphysical brought the physical into place. Therefore, you can forget using physical tools when trying to get to the bottom of it. However, you can derive things that might have happened based on physical observations that we can make today on similar phenomena. For instance, Stephen Hawking, as we've already mentioned, suggests that we can today study black holes in reverse in order to analyze the singularity event that was our creation. Of course, if this is so, it could still only take us back to time *after* t=0; and even then, the black hole itself is not an absolute singularity of nature. What it would do however is help us to get a measure of the distortion of space and time by energy, gravity, and mass. However, we would still not have a handle on the actual singularity itself that brought creation into existence because the events creating the conditions for the singularity to have occurred would have come into existence only from the outer realm beyond space and time, and thus beyond our physical laws of nature. But perhaps we could model the instantiation of space and time, if we knew more about what creates mass, since mass and gravity are tied together (as Einstein found out) and mass is one instantiation of energy; and that is where the God Particle come into play.

But we need to start off with a new postulate straight away: The God Particle is not God! Particles that exist or that we theorize to exist have discreteness in mass or energy, something which God cannot be if he is to be omnipotent, omnipresent, and continuous. You'll remember that this is simply the Trinity argument again revisited. God is not a three-headed or three-person monster, nor is he mush of countless particles surveying the universe looking for opportunities to create mass. But doctrinal theologians and scientists can keep trying; we just can't get away from our insatiable human need to create a visible image of God. From our perspective of being inside the box, God just isn't that visible (unless he chooses to be), and he isn't that imaginable, much as we wish or try for him to be. God is simply far out and far in, and all at the same time; and that's simply another way of saying what St. Paul said (he is through all and in all). When we get to the discussion of hyper dimensionality in other theories, and how it may relate to consciousness and the realm of God, an understanding of this point will be become clearer. But until somebody can measure that which came before creation, we're all just guessing or playing philosophers anyway. Thus you could never be proven wrong by suggesting that the outer realm which God inhabits formed the singularity that is our creation, and indeed brought into existence our physical laws of nature; and this singularity would have given rise to space and time itself. Since those physical laws, by definition, don't exist outside our physical creation, good luck trying to measure his realm or imagine his image! God will be found when he wills to be found. However, that there is a

theorized physical particle that instantiates mass and merits the name
of God probably means it has some special status among physicists,
and special status it has. In order to understand its importance, we
discuss what we know so far about the mass in our universe, and the
implications it has for the unification of the forces of nature.

We start first with those silly quarks. Quarks are those particles,
sometimes also called partons, that are the constituent particles bound
together to make up the nucleons (protons and neutrons) inside every
atom of every element in the universe, first theorized by Murray
Gell-Mann in 1964 to exist yet further inside fundamental unstable
sub-atomic particles that could only be produced in accelerator
experiments, but which seemed to take too long to decay. It seems
the name quark came from James Joyce's Finnegan's Wake: "Three
quarks for Muster Mark", when Gell-Mann had a bout of enthusiasm.
The funny thing about these quarks though is that they don't exist in
nature as free particles bouncing around as they please; they seem to
have an incessant need to be bound to each other, and very strongly at
that. But in 2000, researchers at the Brookhaven National Laboratory's
Relativistic Heavy Ion Collider on Long Island, New York, finally
found something similar to a heretofore unable to produce quark-
gluon plasma, which existed as free particles for only microseconds
after creation, before the quarks had started to bind together. They
also found that this plasma appears as a liquid soup...yum. The
gluons are particles that carry the strong force that keep the quarks
"glued" together to form protons and neutrons; and good thing too,
you and everything around you would dissolve quickly back into this
quark-gluon liquid soup if those gluons lost their power. The only
difference during the very early universe was that the universe was
just too hot for gluons to be effective at stopping the hot energetic
quarks from dancing around each other freely in a puddle. Before
this period, back to the period of inflation, quarks would have been
busy annihilating with other antiquarks to form photons (particles
of light). Only another funny thing would have been happening all
along, as if we needed only one; there were a surplus of quarks over
antiquarks and electrons over positrons being produced, to the tune
of one per billion pairs. And good thing too, the matter that makes

up our bodies and the rest of the universe depended on it; otherwise, the entire universe would just simply have annihilated itself. If God had done that, "let there be light" would have become "let there be nothing but light".

Of course, other interesting things would have also been happening. All the fundamental forces of nature began separating from each other; gravitation would have been first, as it separated from the electro-nuclear force (that still contained electroweak and strong forces together) at 10^{-43} second after creation. Then came the strong force to separate from the electroweak force at about 10^{-36} second, directly before inflation of the universe, which would have occurred at 10^{-35} second. Then "much later", long after inflation, around one trillionth of a second after creation, the electroweak force would have broken up into the electromagnetic charged force and the weak force responsible for certain kinds of radioactive decay. Today it is still not understood how gravity was connected to the other forces because of its relative extremely weak coupling strength; later we'll discuss how physicists are trying to understand this problem and how this coupling may provide a natural link to the ultimate symmetry of the universe, the Theory of Everything, and even our own consciousness, before everything came undone in the imperfect universe.

When the universe cooled down enough to allow the formation of the proton and neutron (bound states of quark triplets if you really want to know, long after the final separation of all the forces of nature from each other), at about a microsecond or so after initial creation, the universe seemingly took a break before continuing with other interesting events, as if God had needed a coffee break. For instance, it would be an enormous amount of time, 3 minutes after creation, when helium and deuteron nuclei (hydrogen with a neutron) would begin to form; thus protons and neutrons were starting to bind together themselves. But the first atoms, which would include electrons bound to the nuclei, would not appear for another 10,000 years! It wasn't understood until 1913 however why the atom was a stable animal; for instance, why didn't the negative electron just collapse in to the positive nucleus? The Danish physicist Neils Bohr would provide the answer; he postulated that electrons inhabit discreet quantum states of angular momentum when they are trapped and circulate about a nucleus such that they can be stable and only jump from one state to another with a discreet amount of energy proportional to integral multiples of the Planck constant, otherwise they remain in their stable quantized state. Thus hydrogen and helium atoms could form, which is what the very first stars in our cosmos consist of. Hydrogen gas, the seed for the stars, started to clump together at about 300,000 years, and the first stars appeared at about 300 million years after creation. Galaxy matter started to group, due to their own gravitational attraction, at about 1 million years; and in 2004, the earliest known galaxy, complete with star clusters, was discovered as having existed already at just 470 million years after creation (Pello et al., ***ISAAC/VLT observations of a lensed galaxy at z=10.0***, Astronomy and Astrophysics, V416, page L35, 2004). Stars in our own Milky Way galaxy only started form at about 200 to 300 million years after creation, while our own solar system didn't come on to the scene until a bright old age of 9 billion years after creation, which means that our sun and earth (and ourselves!) are by-products of an earlier gaseous star cluster arrangement in the cosmos!

Now the link between the strong force and electroweak force is still not very well understood; theories of Grand Unification (called GUT theories), first proposed by physicists Sheldon Glashow and Howard Georgi in 1973 have tried to make this connection. The problem in trying to test these theories is that the force where this unification can be achieved can only be tested at energies of 10^{24} electron volts, an energy that was only present at the time of the big bang, and is the *square* of the best energy that can be produced by the world's most powerful accelerators that man can build today. But we do now have an understanding of the link, at least, between the electromagnetic and weak forces; in 1979, Steven Weinberg, Abdus Salam and Sheldon Glashow received the Nobel Prize for having put forward a theory, first proposed in 1967, that the electromagnetic and weak forces were the result of a single interaction, known as the electroweak interaction that is carried by photons (electromagnetic part) and 3 supposed heavy bosons, known as the W+, W-, and Z_0

(weak part) when the interaction energy is above 100 billion electron volts. Below that energy, the interaction would break down between its electromagnetic and weak constituent parts. At the time of the hypothesis, the three heavy bosons had not been found, although we had already known about the photon.

But in 1983, the Director-General of CERN, Carlo Rubbia led an experiment there that actually discovered these bosons; this discovery not only confirmed the model of electroweak theory but also validated quantum chromodynamics, the study of the strong force between gluons and quarks inside nuclei, on top of the electroweak model. And as a result, Carlo Rubbia received the Nobel Prize in 1984 for this fantastic discovery and his direction of the proton-antiproton collider that would make its discovery possible, along with Simon van der Meer who invented the stochastic cooling technology vital to the collider's operation. CERN thus had the world's first and premier accelerator laboratory and antiproton production and storage facility. Thus, the fierce transatlantic competition was to be engaged with Fermilab in the United States, as it finally had its own proton-antiproton collider and storage ring in place and ready to run by 1987, the year coincidentally that my experimental collaboration was ready to take what would become my first set of doctoral thesis research data, as a Northwestern University graduate student working out at Fermilab. The American joke was that we had a worthy European "adversary" in Carlo Rubbia, as he was as charismatic a leader as he was brilliant; but he also was a good colleague to many physicists in the United States, including my own thesis advisor, Martin Block.

Now being temporarily behind Europe in particle physics for world prestige, discovery, and technology brought much discomfort and angst to the American side of the pond of this collegial rivalry; even the Director at Fermilab, Leon Lederman, didn't even have a Nobel Prize yet – what was a laboratory to do! Well, that angst would indeed prove temporary. In 1987, Fermilab had its first successful run of its proton-antiproton atom smasher, producing center of mass collisions at 1.8 *trillion* electron volts in the center of mass; and Lederman, as we mentioned before, would receive his Nobel Prize in 1988, for work done much earlier during his Brookhaven laboratory days in discovering the muon neutrino. But his success at Fermilab with the successful first operation of the Fermilab collider certainly did not hurt his chances any with the Nobel committee. By 1995, Fermilab had its first major triumph, as the huge CDF collaboration announced the discovery of the sixth and final quark in the standard model of physics, called the top (yes, there is a bottom also). Both CERN and Fermilab had been looking feverishly for it; and on that discovery, Fermilab won the battle. As for the experimental collaboration that I belonged to, we were much smaller and had the modest goal to find the total interaction cross section of protons and antiprotons smashing into each other at 1.8 trillion electron volts in the center of mass, by piggybacking off the particle beam being given mostly to the giant CDF experiment, and located at the diametric opposite end of the accelerator; we also eventually succeeded in measuring this cross section. Two years after our first data run of the experiment, I finished my thesis, using the data we took from this first run and subsequent run a year later that allowed us to measure this proton-antiproton cross section at Fermilab.

Of course, as physicists, we make our share of mistakes along the way, regardless where we are. But none was worse than the personal display of idiocy I exhibited one hot and muggy day during our experimental run, when I was driving back to my experimental counting house to pick up a data tape in order to take home to analyze. Unfortunately, I left the car engine running in park as I went inside for a moment to pick up the tape. The car then decided to drift slowly down the road by itself. By the time I came outside, there it was, with me in hot pursuit, only getting there too late to watch it helplessly roll down an embankment into the Fermilab accelerator cooling pond, with my laundry still in the back seat. After I had the car towed back to the garage to be fixed, stories about my prowess had started to leak out. Before you knew it, I was Captain Nemo. And at that time, there was no Cedric the Entertainer, only me, Cedric the Numbskull. Years later, when I went to a conference at CERN in Geneva, I was introduced to some colleagues who didn't know me; but when somebody brought up the car incident, all I could hear was…"So you're the guy!" But what they didn't know was that every time I returned a rental car to the local motor pool after that, the inspector would look over the car, inside and out, from top to bottom; afterward he would check the trunk. One day I gained enough gumption to ask why he did this; he mumbled something with a little grin on his face that sounded very much like "no dead bodies"! Scientists, like anyone, can be very competitive, they can be very intelligent, they can be very funny, and yes, they can be very dumb, as I can attest.

In fairness, inexperienced young scientists in experimental physics face all kinds of pressures; but in particle physics, the problem can be particularly acute when one joins a large profile and heavily financed experiment with a huge collaboration that takes a large number of years from conception to completion. The story is told of one graduate student with a high profile thesis advisor who was working the salt mines of Utah looking for neutrinos. After slaving away for many years there, the advisor dropped by to see how things were going. The student thought that by then it was maybe time to ask if he should consider writing his thesis. When he approached him, the thesis advisor turned around to look at his student and asked "Who are you?"

The postdoctoral associates are also not immune to pressure and ignorance. For instance, one such poor soul was asked one day to unwind a 1000 foot barrel of thick cable at an access point above ground down to the detector floor 20 feet below the ground. The receiver of the cable on the detector floor was a postdoctoral associate who was supposed to yank it down along the beamline to its destination. When the technician above ground ran out of cable, he got suspicious and went down to the detector floor to see what had happened, because there was supposed to be enough cable above ground to go into the experiment counting room, which I think, the postdoctoral associate would have known; as the technician approached him, he saw a pile of cable sitting in a twisted hump at his feet. Can you say, do it again, Sam? Then there was the hardworking postdoctoral associate who was so accident prone that when he finally had a chance to go to Aruba for his honeymoon, a hurricane dropped in. Of course there are stories of other "postdocs", as we call them, who are often responsible for detectors which they take years to build, but decide on a particular day for one reason or another to reverse polarity and short out the electronics, or let pressurized gas blow the top off their detectors. Yes, stuff happens!

And astronomy students are not immune to this sort of thing. The story is told of one such student who was wrapped up one cold January night with his eye glued to the telescope observing some star. Only problem was that it was so cold that night, his eye *was in fact glued* to the eyepiece because of the frozen condensation from his eye to the eyepiece. And of course, there is the story of the student who found himself one night hanging for dear life from a huge telescope once owned by the Chicago Astronomical Society in 1889. He started swinging helplessly around in mid air as the telescope started to spin on its axis, because he had mistakenly taken off the weights that keep the eyepiece side anchored to the floor. He had gotten into this pickle in order to save the telescope's featured 18.5 inch aperature lens from falling, a lens which had survived the Great Fire of Chicago in 1871 and had been the largest lens in the world from 1864 to 1868. Don't ask me why these things happen; sometimes you just don't know what goes through the mind of a pressured graduate student when he tries to solve a problem. Thankfully, the student and the lens were rescued to safety, but not before some time had passed before another student just happened to drop by that night to do some research.

Unfortunately, young scientists are not the only ones who sometimes still need a tune-up with experience. In 1993, the U.S. Congress in all of its wisdom decided to cancel the Superconducting Super Collider project in Texas, a result that in the near term has eroded the ongoing collegial competition between Fermilab (USA) and CERN (Europe), due somewhat to the scope of the efforts and finances needed to keep up the accelerator race. Of course, the fire ants are now in Texas heaven because there exists this big hole in the ground somewhere in the middle of the state, the size of which can only be seemingly outdone by the size of the state itself. The long term consequences of this decision are yet to be felt; but in the short term, the Europeans will now win the particle physics race, certainly through the early part of the 21st century, with the switching on of power to the Large Hadron Collider (LHC) at CERN in 2007, providing energies that Fermilab will no longer be able to compete with, and now with a clear shot at finding the God Particle. The next generation of elite American pioneering particle physicists will have no choice but to leave for Europe to do their research. The sorry part of this story is that it was the elite European scientists of the 20th century (e.g. Einstein, Fermi) who originally came to America that made American science the best in the world during the past century, and which in large part led to the victories of Allied forces in World War II, for instance. In particle physics at least, that's all about to change; and the impact of this decision on national security issues affecting the United States will have long term consequences yet to be known. And you'll be able to look back and thank the 1993 U.S. Congress for that.

And if you think the situation has improved since then, you should take note of the 2005 U.S. Congress, for instance, who cut $105 million from the 2005 budget for the National Science Foundation from what they had even in 2004. It is hard to conceive how such an endeared institution that was so willing to hand a war hungry President $200 billion over a two year period from 2003 to 2005 so that he could lob bombs at Iraq, then listen to false intelligence that was drummed up in part by some character named Curveball no less, and then be so surprised to find the mess that the United States has been faced with in the aftermath, could then be so shortsighted to have ignored the advice of Nobel laureates when Congress decided to cut off the few billion dollars it would have required to maintain proper research facilities for research in particle physics, basic science, and cosmology research (e.g. SSC, NSF, NASA) for decades to come. But of course, it is the same Congress that also decided not to approve the funding required to repair the levee system that would have kept New Orleans from falling into Lake Pontchartrain. Some say that the money spent on the war in Iraq was necessary for the security of the country; yet I can think of no greater threat to the long term security of a nation than that which allows a country to ignore its own infrastructure or lets its own scientists drift away into other lands where they can do their research. Turn back the clock to the 1930s, and perhaps Germany's Albert Einstein and Italy's Enrico Fermi might have had something to say about that, as they were landing on the welcome shores of the United States of America.

In the long range future, after the LHC, physicists will continue to search for higher energy tools to do their research (LHC reaches a center of mass energy of 30 trillion electron volts); so some jokers in the physics community have begun suggesting the building of a new collider in space, like one to surround the earth (at least it would be round – well almost round) in order to gain far more energy. Remember that in order to reach a true unification force energy, a trillion *squared* electron volt energy would be required. Somehow though, I don't think there are enough governments in the world ready or able to fund that one. If there really is any hope to continue probing a true unification force of nature, it may lie with charged particle cosmic rays and neutrinos themselves. The good news is that they crisscross our galaxy with tremendous energy and near light speed; the bad news is that there really aren't enough of them that want to interact at any focal point (where a detector is installed) to collect statistically significant data with good resolution.

The cosmic rays and neutrinos are so very interesting because they interact with matter quite often through the weak force (called weak because of the very weak coupling strength associated with it, compared to strong or electromagnetic forces). This weak force has become very interesting to physicists during the later half of the 20th century, and ongoing to the present, as interesting kinds of particles called leptons – different from the quarks – have been discovered (leading to Lederman's Nobel Prize for instance); but has really had the interest of physicists are the asymmetric weak force decay modes that have been discovered also. Leptons were also around right after creation with the quarks, and consist of electrons, particles called muons (known to exist since their discovery in 1937 in a cloud detection chamber) and taus (which were discovered at the Stanford Linear Accelerator in 1974), along with the interesting particle that goes with each called the neutrino; and each lepton has an antiparticle mate somewhere in the universe. The neutrino is fascinating, because for a long time, we didn't think it had any mass; but it certainly has plenty of energy to bound about the cosmos at relativistic speeds. The sun for instance releases a lot of them due to the mostly hydrogen fusion reactions that occur in the sun resulting in the release of zillions of neutrinos. Neutrinos come in three flavors associated with the three types of leptons (electron, muon, tau – the tau neutrino being discovered at Fermilab in 2000). They're interesting because they seem to oscillate between flavors, they go missing quite often (for instance the sun should be releasing much more than we seem to see or expect), they interact by the weak force, and they may hold a partial key for many cosmologists in trying to find where all that dark matter is supposed to be in the cosmos, although the latest limits on the concentration of neutrinos in the universe suggest that they cannot make up more than 50% the missing dark matter. But as we shall see, they may have something to do with something else: the matter asymmetry we have in the universe; and that could go a long way toward explaining the source of the fundamental separation of the known forces in the universe, including perhaps even gravitation at the quantum level in the earliest stage of creation when gravity firstly decoupled from the other fundamental forces of nature.

I was involved with one experimental collaboration at Brookhaven National Laboratory in 1990, looking for rare decay modes with the weak force, and even standard model violating decay modes, as we examined the decay of the kaon meson particle via the weak force. We didn't find anything unusual in our experiment except to measure a decrease the upper limit on the possible decay mode we were looking for. However, it did make for a nice radiating muck of photons, leptons, and yes, neutrinos all over the place; in our case though, we didn't have a suitable detector for resolving neutrinos, since we were hoping to find neutrino-less decay modes. Too bad we were unsuccessful; but had we done so, you might have heard about it by now, because the standard model of physics would have been in need of repair!

So what does all this have to do with "The God Particle"? Well, physicists had never been able to come up with an explanation for the construction of mass in the standard model of physics. We have had some success in understanding the constituent elementary particles of nature, the forces of nature, their force carrier particles, and how some forces broke apart from each other quickly after creation from a state of pure symmetry. A natural symmetry breaking of the unified forces of nature along with fundamental particles of nature, which now appear as separate and distinct forces and particles, is implied by the abundance of mass over antimass in our universe and in the distinct and different behavior of these forces and particles; but we haven't still been able to put the whole picture together to understand how they were all once linked with each other. We can explain the connection of the weak and electromagnetic forces to each other, and have some idea how the strong force would have been connected, if only we had the ability to produce that energy required to see it (10^{24} electron volts doesn't come easily!) using GUT theories, for instance; but how do we add gravity and mass to the equation that would allow us to complete the highly prized Theory of Everything? And from an understanding of this theory, there would surely be ramifications for theology.

As for the quarks, antiquarks, positrons and electrons that would have existed (and probably playing musical chair identity change games with each other) at least since the moment before which Leon Lederman suggested we would all be guessing at what happened, physicists have pondered the fundamental question about just what it is that gives mass its energy as opposed to the massless energy of a photon. Mass is that thing that gives us weight, after all; and while there may be too much of it around our own bellies, there still seems to be too little of it in the cosmos, at least if inflationary theory is to be believed! Since we know now about the division of the fundamental forces of nature, specifically the strong force that keeps nuclei bound together, the electromagnetic force that keeps atoms bound together, the weak force that allows particles to decay, and gravity that at the least counteracts mass by holding not only our weight together but the universe itself together, what about mass itself?

Enter Peter Higgs of Edinburgh University. In 1970 he postulated that there was a boson (physicists actually use the name boson to refer to all particles with integral spin which can exist simultaneously in the same quantized energy state as any other boson; if they can't, then they have non-integral spin and are called fermions), that creates a field which permeates the universe and through which all particles must pass in order to achieve their mass. If they go against the grain of the field, they are rewarded with mass; if they flow with the field, they project zero mass. The "going against the grain" concept is what causes (what physicists like to call) spontaneous symmetry breaking, just as what would have happened to the fundamental forces of nature when they suddenly split apart and when there became a surplus of matter over antimatter in the early universe in the period after inflation. Physicists sometimes call this boson "The God Particle" because of the way it permeates all matter in the universe. So there you have it; God is supposed to be a boson. No wonder that church and science have trouble getting along!

If you're not a physicist, about now you're probably feeling a bit disappointed for a moment that there wasn't something more profound or exotic about this particle. If it makes you feel any better, the guy who finds it will probably end up winning a Nobel Prize for someone, and Peter Higgs would be a likely candidate. The thing about this whole naming convention is that scientists usually like to assign names to new discoveries based on the person who found them or theorized their existence. Since a "Higgs Boson" doesn't seem like a profound enough name (for physicists anyway) to give mass to the universe, it seemed prudent to assign an awesome name to it, albeit in an almost sarcastic or cynical way. So we call it "The God Particle". We can admit at least that the significance of the name gives respect to the particle!

The same phenomenon occurred with naming the big bang "the big bang". In the 1940s, when George Gamow had presupposed the existence of a very hot dense early state of the universe at creation, with conditions that would be similar to that which occur in atomic bombs that use thermonuclear reactions, not many people bought into his theory that there should be a remnant of the big bang still remaining today, in the form of cooled high energy photons left over and still lurking about in our universe. British prominent astronomer Sir Fred Hoyle (1915-2001), well-known for his work on stellar nucleosynthesis, sarcastically referred to the creation theory as a "big bang". That was another way of saying he thought Gamow's idea was nuts, because he thought at the time the universe had no beginning. Then in 1965, Arno Penzias and Robert Wilson of Bell Telephone Laboratories made their discovery of the cosmic microwave background that proved that Gamow was right all along. But Hoyle continued with his own creative ideas and showed interest in the idea that life had a cosmic origin, and not an earthly one. He had taken an interest in the "little green men" results of his Cambridge colleagues, Jocelyn Bell and Anthony Hewitt, who did an experiment in 1967 that led to the discovery of pulsars (and not "little green men", but more on that later); we know about that because Bell was a graduate student at the time who was awestruck at the sight of Hoyle showing up for her talk about pulsars, which her advisor and she had discovered.

Now if this Higgs Boson were found, many physicists would be very happy, because they have been incorporating it for many years already into the standard model of physics. If they don't find it, many of those theories would need rewriting, and a call for new physics would go out to the world. For the rest of us, we tend to contemplate nature and then accept what is. We don't think much about those pointlike quarks combining with each other via a gluon field carrying the strong force to form protons and neutrons which are attracted to various nuclei, or those pointlike electrons combining with nuclei via a photon field carrying the electromagnetic force to form atoms and molecules. So it goes; but the game never really ends for physicists. For instance, the quarks and electrons themselves have mass; the top quark itself was just discovered by the CDF collaboration at Fermilab, when I was working there (at the diametric opposite end of the ring) no less. By the way, there are six quarks in all; and you'll love these names: top, bottom, strange, charm, up, and down. What physicists lack with their imagination they seem to make up with their humor! Yet who is to say quarks and electrons don't have structure; and if they do, then there is a whole new generation of particles underneath that cover yet to be found. And the Higgs Boson, excuse me, the God Particle itself has mass. The best estimate of its mass equivalent energy is something above 117 GeV (giga electron volts);

but theoretically it could go as high as 251 GeV. So it goes with the standard model of physics.

In physics however, when assigning masses, you find that the sum of a particle's mass usually is much different than the sum of its parts! For instance, the proton is a bound mixture of two real up quarks and a real down quark; the proton's mass is about 938 MeV (mega electron-volts), the up quark mass is 5 MeV, and the down quark mass is 10 MeV. So what happened to give the proton the other 918 MeV of its mass? The answer lies in the binding (potential) energy given to the quarks as they bind to each other inside the proton. And then there are all those virtual "sea" quarks that seem to raise their ugly head for a moment and then annihilate and disappear again constantly inside the proton. Now the three (up, up, and down) main valence quarks (as they're called) don't just sit there, even if they don't disappear. They're constantly moving around, held together by what seems like an invisible elastic string. The further apart they get, the stronger the tendency for them to snap back toward each other. Think of the proton on a *huge* football field of length one millionth of one nanometer (10^{-15} meter); the quarks would then be like three football receivers constantly dancing past each other like yo-yos on a string. What happens is that the gluons, which have no mass, do the dirty work by keeping the quarks in a "sticky" mood by giving them potential energy as they stray apart, and even taking some of that binding energy away as they get too close. It seems these quarks like each other, but not that much; or perhaps the gluons just get too jealous! In any case, the whole process contributes to the mass of the proton and any particle with quark-gluon structure. The Higgs particle does the same thing with the quark, for instance. Funny thing with the Higgs though; it also interacts with itself to give itself mass; and I won't even try to analogize that one!

Fortunately, there are other theories, outside the standard model of physics that are under development which actually have no need for a Higgs boson, and allow particles to obtain mass in other exotic ways. That's right; there are theorists out there trying to find an alternative to God being a particle! One intriguing theory originally put forward by German theoretical physicist Burkhard Heim (1925-

2001) introduces something called a metron, an elementary surface element which creates mass and inertia by interacting with other metrons. The original theory is 6-dimensional and is done within the framework of a Unified Field Theory, a Theory of Everything which seeks to combine quantum mechanics with General Relativity in order understand all forces of nature from a common symmetrical beginning and viewpoint, concepts that Heim really initiated. This idea is analogous with what present string theory tries to do because the metrons are effectively the same objects that string theory uses to describe multidimensional space; it too tries to find the Theory of Everything. But what some establishment physicists don't really want you to know is that Heim's theory has correctly predicted masses for many fundamental particles (including leptons like the electron and baryons like the proton) to within 0.1 percent, something no other theoretical model has yet been able to do; its predictions for excited energy states of atoms and something called the fine structure constant have also been "on the money". The reason you don't hear much about him or his theory is that Heim stayed out of academia and never bothered to publish, except for one article he gave into publishing; and hence critics argue that his theories were never held up to the same rigorous scrutiny that everyone else had to obtain. Part of the explanation lies with the disability of Heim, who suffered loss of eyesight, hearing, and his arms to an explosion when he was 19; hence he became a recluse and indulged all his time into Einstein's relativity theories, never bothering to publish until he was 50 years working on it! Although even denounced by many theoretical physicists, Heim produced a mass formula that is his theory's greatest secret attraction; it has been extremely accurate with fundamental particles, certainly to within the error of measurements. The theory has since been extended to 8 dimensions in order to explain particle interactions, and to allow for the derivation of the particle mass. But here is where it gets very interesting, because it uses the geometry of spacetime to explain not only gravity, as is the case with General Relativity, but all forces of nature; it is through this multidimensional geometry that all particle mass would come about! More recently, it has been further extended to 12 dimensions as a basis to explain more features of gravity, such as anti-gravity, that is supposed to be responsible for the expansion of the universe, and a force (as of yet unobserved) to convert electromagnetic fields into gravitational fields. Theories of quantum gravity today make good use of these concepts; even string theory, while officially independent, makes use of much of the multidimensional mathematical geometry that is used in Heim's theory.

The reason multidimensional theories have modern day theorists' attention is that they provide a way for explaining mass and how it might warp our four dimensional universe of space and time. Perhaps one way of understanding it, as Heim's theory seems to suggest, is that mass causes a warping along an orthogonal dimension to the dimensions of space and time; and this warped dimension pushes or puts pressure on the dimensions of space and time that is the cause of their observed warp. The greater the mass, the greater the warp, and so on. The idea is not so far fetched from a quantum mechanics, because Heisenberg Uncertainty Principle leads one to an equation that links a particle's rest energy (its mass) with the inverse of its effective quantum mechanical size. If this size were equivalent to a dimension that becomes projected onto the dimensions of space and time, as we'll see string theory tries to do, one can see how the warping of space and time by mass could happen; the measure of this warping of space and time is nothing other than its gravity.

And then there is the issue whether quarks and electrons (pointlike in the standard model of physics) themselves have structure, i.e. whether they themselves are made up of other particles inside. Physicists usually like to stop with the bosons and fermions as being as far as it goes, such as the pointlike electrons and quarks, which both happen to be fermions; but what they really don't know is if that is as far as it goes in terms of particle structure. You see, we never have been able to drum up powerful enough (or expensive enough) detectors (e.g. accelerator laboratories) to see inside these particles; and that's what is making it so difficult to see even the Higgs boson itself, if it exists. Only the LHC at CERN in Geneva, Switzerland, has the opportunity to observe such structure. Researchers have been gearing up for this search for a long time. But it must be noted that even LHC may not produce enough energy (running up to 30 trillion electron volts in the center of mass) to probe possible electron or quark structure. For a while, it was thought that maybe even the CDF collaboration at Fermilab had seen evidence for this structure inside the quark; but the evidence for it so far has not been very strong.

But that never keeps a physicist from pondering more pie in the sky, such as theories about the WIMPS we discussed earlier, along with the yet undiscovered supersymmetric particles of nature. And they have some names that are real doozies: photino, wino, gluino, selectron, squark, sneutrino, zino, gravitino, and yes let's not forget the higgsino, the supersymmetric pair particle of the Higgs (God) Particle. (The *wino*, you say!) These particles come about in supersymmetric string Theory, a variant string theory, perhaps gone amuck, that says that all particles are just tiny strings with different vibration modes, and suggests there is a boson for each fermion and vice versa. You really don't want to know more about this, except to say it has earned tenure for a whole bunch of people who otherwise couldn't get it by discovering new particles themselves anymore, if for no other reason there is just too much time and expense needed to keep running bigger and more powerful accelerators to keep finding new particles. However, these supposed supersymmetric particles are probably not going to be found any time soon; but any new theory has to be able to predict something if anyone is ever going to believe it, especially one that is a leading candidate for finding the Theory of Everything. But some particle physicists actually think they will find all kinds of goodies at LHC at CERN, in Geneva, Switzerland, as collision energies reach those 30 trillion electron volts in the center of mass, such as some of these supersymmetric particles and maybe even a mini black hole event or two, not to mention the God Particle. While many massive boson particles will undoubtedly be found, none of these particles or events is ever guaranteed, or even likely, to be found. However, with 30 trillion electron volts for European scientists to work with, that ain't Swiss cheese.

What you might be interested to know, however, is that these so far heretofore purely hypothetical particles are part of one theory behind some of that missing mass that the cosmologists say must be in the universe, as predicted by both inflation and relativity theories.

Today some physicists conclude we really do actually only see 4% of the total matter in the universe in order, to explain our constant but stable expansion in the cosmos; 35% of it is supposed to be dark matter (real particles with mass), and the other 60% of it phantom dark energy. This energy you'll recall could be described by that cosmological vacuum energy constant that bothered Einstein so much, some of it could involve the inflationary energy of the universe, and some of it could involve ordinary light. The dark matter is supposed to be a mix of dark hot gaseous particles surrounding the galaxies, neutrinos with some mass, and yes, those supersymmetric particles we just finished talking about. Unlike the supersymmetric particles however, there has been some experimentally observed evidence of dark gaseous matter and neutrinos with mass; but not nearly enough has been found to account for all the missing mass that theorists need to complete their conventional theories. Further scenarios for explaining the rest of the missing dark matter could be with densities of black holes or brown dwarf stars that, by their definition, simply don't emit any light for us to see them. The supposed missing mass and energy in the universe just illustrate however that many scenarios to explain the makeup of our universe still exist, and how much is out there for us yet to learn. What we do know is that all evidence keeps pointing to a generally open universe that is constantly expanding and even accelerating, and one which won't simply collapse back on itself, if that is supposed to be of any relief.

Thus the theological consequences that can be drawn from what we know so far from the particles that have been discovered in particle physics are not earth shattering. Even if the Higgs particle were found, it really would be simply a matter of interpretation what God really had to do with it! The bigger impact however for theology is what we have learned about the spontaneous symmetry breaking of the forces of nature, and what is theorized to create mass, because it helps to verify the existence of a finite imperfect universe that was once in a perfect state. And it corroborates the cosmological theories that the universe had begun from a distinct initial singularity that doesn't simply oscillate back onto itself over time; that's a nice way of saying we're not in for a Big Crunch one day, say 10 billion years or so, plus or minus an eon or two, not that anyone really cares today about it. Then, unlike theories on the evolution of life, scientists and theologians can actually agree on something: that something or someone initiated the big bang scenario. As it is, if physicists are interested more in the details, as they should be, theologians and churchmen are more interested in the final result. Leon Lederman in his book **The God Particle** asked the question, "If the universe is the answer, what is the question?" Theologians and churchmen might respond with an equally tongue and cheek response: evidence and proof of the existence of God. Scientists might have a different answer, albeit I would suspect that they have a more difficult time trying to drum it up. And once again, we come back to the point that scientists can't seem to beat, yet even vainly try to describe: just how did that singularity of our existence come into being in the first place.

The answer must lie somewhere in the outer realm of God which would encompass but yet is beyond our universe of physics, space, and time; and that will frustrate physicists to no end (and to the delight of people of faith) because there is nothing out there they can get their hands on for measuring or observing outside of space and time. However, there is still hope for scientists; and it comes with multidimensional theories of creation such as Heim Theory and the analogous string theory. If and when predictions which are implied with these theories are proven to be correct (such as mass that can already be predicted for fundamental particles in Heim Theory), we can gain more confidence in the understanding of how our universe works and how it was created, which include mass and the initial unification of all forces of nature, and their resulting breakdown. And there is no reason to believe that God couldn't create parallel universes with multiple singularities from within his outer realm. And that idea probably brightens physicists' curiosity (while it causes shoulders to shrug for people of faith). Historical theological distaste for Giordano Bruno and the exuberance of the string theorists would be proof of that argument; although in Bruno's case, his ideas about multiple Messiahs and crucifixions were probably meant more as sarcastic jolly than they were profound philosophy, while string theory is a serious attempt to find the ultimate Theory of Everything. Yet string theory may lead in a direction where even scientists were not expecting, as Burkhard Heim in later life started to find with his own multidimensional ideas; dimensions of thought, consciousness, and spirit started to creep into his math from which some physicists today probably have more amusement with than seriousness. Yet models of consciousness and multidimensional space are today independently developed, and by different professions. And there is a strange curious similarity that should get some theologians to wake up and take notice. As we'll see, they may find that there is more than just math behind those equations. And the exciting part of it is that scientists, psychologists, and theologians may not yet be aware of the surprises these models may bring to the table of the Theory of Everything. But first, there are other clues that have been left behind in our universe which we should pursue that may also help us gain more insight into God's original plan for our universe, and indeed may help to answer other questions, such as whether there really is other life out there, what other universes could have formed, and whatever else he may have been up to. And you might guess that they are connected to the question of symmetry in the universe, specifically space, time, and charge/mass symmetry, and the lack thereof. For this, we once again solicit the help of those particle physicists, who seem to have a nose for this kind of thing.

X. The CPT Conundrum

When we talk about symmetry breaking, what do we mean? If something has symmetry, then any interaction that uses it must behave the same way in reverse, whether we are talking about matter to antimatter, time to backward time, or space to spatial inversion. When we talk about symmetry breaking, (or spontaneous symmetry breaking as physicists like to call it) then we are talking about something that happened along the way to cause a sudden irreversible loss of symmetry. In the case of the big bang, and our fundamental forces of nature, we believe that when the universe was at its singular t=0 point, the universe was a very, very hot singular point in space; but it existed in a moment of perfect symmetry. However, as the clock started, the universe started an immediate and expansive cool down that led to the loss of symmetry between forces and elementary particles of nature. Thus what used to be one force, suddenly became four, and what was just one particle oscillating with itself became an initial host of quarks electrons, and their corresponding antiparticles. Perhaps you can see this by example. When a distant traffic cam takes a quick snapshot of a line of cars going swiftly down a freeway, it appears as a streak of light on the emulsion. However, as the cars slow down, or as the camera moves in or improves its resolution, the individual cars that were once part of one singular streak become discreet and different. The same thing happens with spontaneous symmetry breaking, except that something needs to trigger the process.

Now imagine a two dimensional diagram of space and time, where each axis has a positive and a negative pole. In the world of quantum mechanics, it is possible for something to seemingly come out of nothing where symmetry is broken, only to necessarily recombine again at some distant point in order to maintain a conservation of nothingness. If the story of creation started out from an infinite vacuum where this space and time diagram was nothing more than a singular point to start with (space and time didn't exist yet), it is possible then in quantum mechanics for something to tunnel through a virtual singularity point onto some real point on a space and time diagram, but only if it has a corresponding anti-something that exists somewhere in virtual or real space and time.

One of the fundamental principles behind inflation theory since its conception was that this nothingness is actually conserved in some way. Only our universe as we know it broke off along only the positive side (for convention) of the timeline axis, leaving the negative timeline behind; this by itself leads to the distinct and interesting possibility about what is actually happening on the other side of the time axis in a negative time running universe! For instance, did you ever wonder why time in our universe can only seemingly tick in one direction? Yes, of course, we can slow it down; we can twist it or warp it, depending on the forces acting upon it and the speed of any given reference frame. Yet the universe seems split by a positive timeline we all seem to know about, and a negative timeline that we seemingly know absolutely nothing about. But in the earliest moments of creation, there could have been two universes (one along the positive timeline and the other along the negative) right next to each other forming from a common reference point; thus it may have been possible through quantum mechanics for a brief moment of time when gravity could act on particle/antiparticle pairs (still in equal numbers, but able to appear and disappear at random while maintaining the principle and conservation of nothingness) that were straying along each side of the time axis. Only gravity might have been fast enough to operate on both sides of the timeline horizon after it broke off from the otherwise still unified force of nature, since only gravity could have interacted at speeds greater than light speed in the real universe, if only for an instant. (Circumstantial evidence for gravity's quickness is described below.) The brief interaction by gravity across the timeline could have been responsible for inflation, resulting in a repulsive gravitational force acting on the particle/antiparticle soup on each side of the timeline axis. Inflation theorists use the language of imaginary time instead of negative time in their description of creation; imaginary time would thus be time proceeding along a negative timeline with respect to our positive universe timeline. Also, physicists have not yet developed a complete quantized gravitational force field theory, except that they make use of a mediator particle of the field, called the graviton. Now once particles had long separated across the timeline, the gravitational interaction could only be attractive for particles and antiparticles in its respective timeline, thereby bringing an end to the brief inflationary period of the universe. Along the negative time axis, all the particles that were inside it would have broken off from our universe into a thought provoking negative timeline universe. However, in just a brief moment of time after creation, when gravity separated off from the other still unified fundamental forces of nature (electromagnetic, weak, and strong), the first spontaneous symmetry breaking would have occurred that would lead us down a permanent road to imperfect asymmetry and finitude in our universe.

It turns out that observations of our universe are actually consistent with a gravitational interaction rate which could have been as much as 10^{10} greater than our *present day* speed of light in the early universe. For instance, how is it that a black hole can react with its surroundings if the definition of a black hole is that something going the speed of light itself cannot escape it? And how do galaxies that are millions of light years apart manage to gravitate about each other continuously like the moon orbiting the earth, and so forth? Obviously gravity has no problem affecting its surroundings from great distances that seemingly defy the speed of light; thus primordial gravity's warping of space and time should have occurred with an interaction speed that is much higher than today's light speed. However, it needs to be pointed out that Einstein's General Theory of Relativity predicts that gravity must have the same speed as that of light, but under the same conditions in space and time. Measurements by the National Radio Astronomy Observatory (NRAO) in Charlottesville, Virgina, in 2003 have tried to verify this prediction by measuring variations in the light image from distant quasars bent by jupiter's gravity; they obtained a speed of gravity number consistent with the speed of light, with a 20 percent error in measurement resolution. Unfortunately, this experiment is the subject of much controversy about the proper way to measure the speed of gravity, because, by definition, their experiment was dependent on the speed of light to obtain the quasar light data here on earth!

And of course, maybe the most interesting cosmic question of all can be asked: just how did those distant galaxies get out there so far so fast? The furthest galaxy to date was discovered in 2004 by the Hubble Space Telescope to be 13.2 billion light years from earth. Since we know that galaxies take about 500 million years to form, that would have given this particular galaxy, filled with massive stars, no time to get to where it is, in a universe that is 13.7 billions of age! Given the causal relationship of matter between here and there, it would indicate that something in the universe must be able to travel that far, and very, very fast. And we still haven't reached the peak resolution that will allow us to see out to the visible universe horizon; that will have to wait till at least another decade with the launch of the James Web Space Telescope. As we see below, a primordial interaction rate of 10^{18} meters per second, 10^{10} above the speed of light, could answer some of these questions and also shed light on the theory of inflation itself; at this speed, however, one could go from earth to the visible universe horizon (10^{26} meters) in just one year! That's moving! But who ever said, God wasn't fast!

Also, a higher gravitational interaction rate than today's speed of light, produced by gravity's spontaneous symmetry breaking after the initial moment of creation, would have allowed quarks and electrons to interact by gravity before the rest of the forces of nature (still unified) could do so. Thus, we would find that inflation could have resulted from this interaction by gravity before other forces would have had a chance to stop it. For instance, we can naively consider that quarks inside a proton have one thousandth the size of the proton itself (by virtue of the ratio of masses). Thus, if we take the width of a quark and divide by a gravitational interaction rate, that we are theorizing to be 10^{18} meters/second in the early universe, we get a possible interaction time for two quarks to "see" each other of 10^{-36} second, coincidentally about the time inflation is thought to have actually started. In some ways, this idea of a very fast gravitational interaction rate may become a substitute for inflation itself, except that we don't need to assume here that any primordial phase change with gravity had to occur; on the other hand, if we consider that inflation is typically described in units of today's timepieces only, the two ideas may well be quite consistent with each other!

But after inflation, more spontaneous symmetry breaking did occur with the rest of the forces of nature which resulted in mass being produced at a higher rate than anti-mass. How could this happen; after all, if particles and antiparticles were originally in equal numbers, they would have annihilated each other as they found each other, even after inflation? And one of the most fundamental tenants in all of physics, certainly up to the middle part of the 20th century, was the conservation of symmetry in the universe. Neutral particles could only decay into other neutral particles, or else decay into equal numbers of negatively versus positively charged particles. If one decay particle splits off or curls up in one direction, another must split off or curl up in the opposite direction in the center of mass. If some reaction occurred in positive time, its antithesis reaction would occur in negative time. Certainly, test of the electromagnetic and strong force interactions have all convincingly showed that symmetry in the universe is always still maintained. Physicists have even developed a whole laundry list of symmetry laws, laws under which conservation of the quantity being measured is maintained, even after any physical interaction in nature. There is conservation of energy, momentum, angular momentum, charge conjugation, time reversal, and spatial inversion; even particle families have the same number of member particles conserved under an interaction, such as the baryon and lepton families. Certainly, the strong and electromagnetic forces are willing adherents to these laws, most of which were formulated by the 1930s.

So it was naturally assumed that all forces of nature were conserved under these laws and symmetry in the universe was maintained, except for the nagging question of universal mass asymmetry. It certainly sounds intuitive enough, to even the most logically challenged of people, that symmetry is maintained in nature. Just look at your own self in the mirror in the morning and see what you see. For some, this might not be a good idea; however if you notice something asymmetric in your own appearance, you might feel the need to call a doctor. Well the universe seems this way also. The earth certainly seems oval enough, the sun and moon seem round enough, and the orbits of the planets and celestial bodies about each other certainly seem symmetric. Even Werner Heisenberg, author of the Heisenberg Uncertainty Principle, in 1932 postulated that there was nuclear symmetry of all nucleons inside the atom. And by the time we get to 1954, the postulate of charge conjugation, space parity, and time reversal symmetry, taken together as a product, was mathematically deducible and important for quantum field theory. And yet, up to this point, thanks to bias, nobody was even beginning to consider or test the faintest possibility of some force violation of any of these laws, much less gravitation, which seemed so well explained by General Relativity. So everything was peaches and cream, except for mass, right? Well, not so fast.

During April 1956, physicists gathered at the University of Rochester to try to resolve one particular oddball mystery. They had found what seemed like the same particle decaying via the weak force into two different sets of particles, called pions; sometimes the particle would decay into two pions, other times it decayed into three. However, the pion has spatial parity -1, meaning that it has a negative quantum mechanical spatial wave distribution function when space is inverted (+1 means the wave distribution stays the same under spatial inversion); thus the decay reaction sometimes had net parity $(-1)(-1)$ = +1, and other times it had $(-1)(-1)(-1)$ = -1, meaning that the total spatial parity of the reaction (denoted by the P operator in physics) didn't seem to be conserved at times, while other times it was. The problem became known as the theta-tau puzzle.

Enter two young ambitious physicists, Tsung Lee and Chen Yang, who had escaped China for the United States as students during World War II and the Sino-Japanese conflict. They sought out Enrico Fermi, having known about his famous work with neutrons and nuclear chain reactions, and earned their doctorates at the University of Chicago during the 1940s. In the early 1950s they found each other again as fellows at Princeton's Institute for Advanced Study. By the time of the conference in 1956, Lee had moved on to Columbia University while Yang remained at Princeton; but together they delivered a paper to the conference that proposed that elementary particles could come in two forms, a process called parity doubling, meaning that each particle had simultaneous dual quantum states. At this point, they hadn't entertained the thought publicly of some sort of parity symmetry violation going on. However, you might have guessed that there weren't many takers for their parity doubling idea either.

Now attending the conference was another up and coming prominent and charismatic physicist; his name was none other than one Richard Feynman, who became quite famous later in life for his development of the field of quantum electrodynamics (better known as QED, that describes quite precisely charged particle interaction, spin, and energy levels via the electromagnetic force), for which he would eventually share the 1965 Nobel Prize in physics. Feynman would come to be everyone's favorite physicist and lecturer during his lifetime, not only for his natural genius, but for his many books and lectures, his social activities (he took up the drums, writing Chinese, and lock safe mechanisms), and his downright refreshing pleasant personality. It is said that his Ph.D. thesis dissertation on quantum mechanics so impressed his thesis advisor at Princeton, that he presented it to Albert Einstein, also still at Princeton, presumably thinking it would appease Einstein's remaining trepidations with quantum mechanics. Einstein would remain unconvinced. After finishing his doctorate, Feynman would find himself administering the computers and punch card systems in use at Los Alamos during the Manhattan project (he had a keen interest in computing and is said to have first considered the idea of quantum computing), and from there rise to the level of theoretical physics professor at Cornell; yet still he could not find personal fulfillment.

By the time of the Rochester convention in 1956 however, Feynman became professor of physics at CalTech, and was quickly becoming a pre-eminent but yet still not well-known theoretical physicist; but he had also taken an interest in the parity symmetry problem. At the conference, he shared a room with a young experimentalist named one Martin Block, who suggested to Feynman in private that just maybe the parity conservation symmetry of the universe was actually being violated after all, at least with weak force interactions anyway (*"Surely You're Joking, Mr. Feynman!"*, W.W. Norton, 1985). Feynman didn't really believe that, and in fact later confessed that he made a bet with someone that parity wasn't being violated (a bet he eventually lost); but he brought it to the attention of the Chinese physicists, Lee and Yang, especially after hearing their talk about parity doubling, which seemed equally unappealing. Lee and Yang subsequently left the conference rejecting the idea, but a month later decided to give the idea another look by examining all the weak interaction data that was available at the time, just as they were beginning guest scientist appointments at Brookhaven National Laboratory on Long Island, New York, the summer of 1956. They proposed certain successful experiments in radioactive weak decay that would indeed show that parity was in fact being violated after all! In 1957, both Lee and Yang were awarded the Nobel Prize in Physics "for their penetrating investigation of the so-called parity laws". As for the theta and tau particles, they were eventually found to be a unique set of separate quantum mixtures of the neutral kaon meson particle and its antiparticle pair, which had different lifetimes before decay. It wouldn't be until Feynman's book came out many years later when Martin Block's contribution to the whole idea of parity violation, first mentioned at the 1956 Rochester conference, would be publicly known; and it turns out the book was published the same year that I became Martin's thesis student at Northwestern.

But that wouldn't be the end of the picture for those (theta-tau) kaons. It turned out that the weak force that causes them to decay would have other symmetry conservation violating components also.

Princeton researchers Val Fitch and James Cronin in 1964 were studying the neutral kaon meson particle decay modes and showed they seem to have a particular charge *and* parity nonconservation mode when taken together. Ultimately this discovery would also lead to a Nobel Prize for each in 1980. The significance of this result was that if these particles underwent a charge-parity (a CP operator) violation, while charge-parity-time (CPT operator) symmetry taken together still seemed to be conserved, then time itself must be violated also! For the first time, physicists had gained some idea or clue as to where all that missing anti-mass went, versus the existing mass that is in the universe today, since they still had an implicit need to believe in the existence of universal symmetry. (The fact that Val Fitch was once an undergraduate at McGill University in Montreal, where I teach as an adjunct professor, and the fact that James Cronin's parents originally met at Northwestern University, where I did my graduate work, have nothing to do with my making a big deal out of it; but it's nice to mention it anyway!) Their discovery however did turn out to be an infamous finding, which implied that in negative time the time violating weak force component would produce a surplus of antiparticles if the conservation of charge, space parity, and time (CPT) was still to be maintained; namely if you reversed time for all reactions and decays under study, any previous particle/antiparticle deficit in the neutral kaon decay system would be accounted for. When the discovery was made, physicists went into a theoretical sizzle in order to finally get a handle on the scintillating question about why our present day universe (in positive time!) has so much matter over antimatter. Thus, they finally had something to work with, in order to answer this fundamental question. It might have also provided a clue as to what could be going on in a hypothetical negative time universe. But the only problem that would remain was that the total amount of time-violating particle decays found in the weak force would not completely explain the full amount of particle over antiparticle asymmetry that we find in our universe today. All theoretical models of particle physics, including quantum field theory, and the standard model have always incorporated the idea of CPT conservation in the universe. Could part of the problem lie there?

Well, in 1986, the Kamiokande collaboration in Japan counted and confirmed a deficit of solar neutrinos in their detector, from what was expected from the sun. Later in 1998, the next generation Super-Kamiokande collaboration found that neutrinos were not only appearing to be in deficit, they were oscillating into other flavors of neutrinos (Super-Kamiokande Collaboration, *Evidence for Oscillation of Atmospheric Neutrinos*, Phys. Rev. Lett. 81, 1562–1567, 1998). Thus, the feature of neutrino oscillations confirmed the view that neutrinos actually have mass. Then in 2001, physicists at the Sudbury Neutrino Observatory (SNO) in Sudbury, Ontario, Canada, helped to put the neutrino mass question "nail in the coffin" when they discovered that the mysterious loss of neutrinos from the sun (only 35% are detected from what is predicted) is due to the fact that some of the electron neutrinos (the only kind coming from the sun) are actually oscillating into other flavors along the way to earth (Physics World, *Solar neutrino puzzle is solved*, July 2001). Thus for the first time, we had found evidence to suggest that a fundamental particle of nature does undergo identity change on occasion, a kind of identity theft with its nearest quantum neighbor particles if you will, something supersymmetric string theory would have required, in the very early universe at least.

In the meantime, the LSND collaboration working in Los Alamos, New Mexico made an astounding discovery. They also found that antineutrinos were oscillating into different flavors, but at a different mixing rate (LSND Collaboration, *Evidence for neutrino oscillation from muon decay at rest*, Phys. Rev. C54 2685-2708, 1996)! What all this meant was that the cherished theorem of CPT conservation was not only in doubt, but probably being violated. In a paper by V. Barger, D. Marfatia and K. Whisnant (*LSND anomaly from CPT violation in four-neutrino models,* Phys. Lett. B 576, 303, 2003), it is argued that the LSND result can be explained in terms of the other existing data with the addition of a fourth "sterile" neutrino (impervious to the weak interaction) and CPT violation. Neutrinos are a funny breed in that they appear to oscillate from one flavor to another (muon, electron, and tau are the known neutrino flavors to exist), implying that they themselves have some mass, which has been very hard to nail down, because neutrinos not only interact by the weak force (very weakly in fact), they tend to gallop across the cosmos passing through earth like it was a lousy fish net. The result is that it is hard to get statistically reliable data from neutrinos; but on the other hand they have very high and relativistic energy, travel great lengths, and have the ability therefore to unlock some secrets of the cosmos itself. Today some particle physicists are scurrying around trying to figure out how to design neutrino factory accelerators that could provide statistical input to a new generation of experiments that could confirm or squash the existing and yet immature neutrino data.

Now it turns out that if the conservation of CPT is really being violated in our universe, another channel would be provided for missing mass production and the resultant universal particle/mass asymmetry abundance, in addition to the CP violation we already see in the neutral kaon meson particle decay system. Invoking Valentinian mythology, you could argue that God, the perfectionist of the ages, created a nonperfect universe that therefore was meant to be imperfect and finite. If one was trying to figure out why God brought this imperfect cosmos into existence in the first place, why God would have allowed this asymmetry to have occurred, and why he did it when he did, one could consider, for instance, the speculation that God needed to do it in order to cast out Lucifer, the devil, in the great epoch battle with rival archangel Michael for control of the heavens; this would imply of course that God must have had a major fit! In theology, serious merit could in fact be given to that very argument, because the devil was cast upon the earth and allowed to cause mayhem and temptation to all that dwell upon it, according to the book of Revelation in the New Testament of the Bible. (As an aside, we do find another battle that is recorded in the book of Jude in the New Testament, where Michael and the devil "duked" it out, yet once again, over the disposition of the body of Moses. One wonders though what the devil would have wanted to do with it!)

Now certainly from a temperature standpoint, it could be understood that God may have had a major fit and been very angry for some reason, and angry indeed! It would have been 10^{32} degrees

Kelvin at the epoch moment when inflation and/or expansion of the heavens would have occurred after the big bang, and after the first symmetry breaking of gravity from the other fundamental forces of nature. That's no less than 10^{25} times the inner core temperature of our present day sun. You talk about getting hot headed; wow, it pays not to anger God! But if God was angry for some reason, his venom would have spilled out of the heavens into other domains. In scientific language, you could say that heat gave way to CPT violating universes; in other words, God had a CPT conundrum.

As for the physical considerations of this violation in our universe however, we could explain abundance of mass over anti-mass, and the need to have a negative timeline universe to compensate for this loss of symmetry. But the interesting thing about that is what we would see differently in a negative universe from our own. CP violation, as found by Fitch and Cronin, would cause a production asymmetry of particles to antiparticles in our universe, counteracted by a corresponding production asymmetry of antiparticles to particles in the negative universe. CPT violation gets slightly more complicated however; it would not only imply possible production asymmetries between the two universes, but it would also imply production rate asymmetries between the two universes, even when the same (oppositely charged) reactions were occurring between the two. However, even more seriously, unlike with CP violation, CPT violation presents physicists with a problem. (Not unlike the Trinity, why do these problems always seem to come in threes!) The beloved dogma of quantum field theory (that tries to describe all force fields in terms of quantum mechanics) is thrown into a tether because CPT conservation is one of its mathematical consequences. Not that you really care about that, unless you are a physics graduate student (because theorists will undoubtedly find a renormalization for it somewhere, such as something called the Standard Model Extension); but it does effect what you would expect to have happen along the negative timeline. Namely, the particle makeup of a negative universe would likely be much different, such that atoms and molecules may not have formed in the same way, or at all, even if they were in the form of anti-matter atoms and molecules. But if it makes you feel any better, CERN in 2002 created the first element in the anti-periodic table, anti-hydrogen, consisting of a positron orbiting around an antiproton nucleus.

Well, this all gets even more interesting because Andrei Linde at Stanford has proposed the idea of budding or new universes that can form out of our own universe if the conditions were right. The same quantum fluctuations would have to occur that gave rise to our universe; but a theoretical argument is made that the laws of physics would allow for it. This goes back to the universe in a bottle argument discussed before; but the idea is that if we put together enough energy and temperature together at a close enough distance, a wormhole could temporarily form that would cause a universe to "spawn off" from our own. Because these wormholes are likely to be unstable, it's not something you would want to cross into or out of very often, you understand. Only problem is that you need an energy source to create the energy density necessary for gravity to be in unison with the other fundamental forces of nature, namely during a moment of perfect symmetry, as would have occurred during the big bang, so that a false vacuum of positive energy could form in a region of unstable space and time (eg. positive and negative space and time which quantum mechanically overlap). That energy required is a whopping 10^{28} electron volts, which is a factor of 10^{16} greater than any particle energy which can be created today! Good thing too, you might say; after all, who wants new baby universes to form without knowing what it might do to our own! It turns out even a flea has that kind of mass equivalent energy; but the difference is that a flea doesn't concentrate his mass over 10^{-35} meter, which is the focal size necessary to destabilize space, as we'll see why later. Now a flea is small, but he isn't that small! So you can relax, because if a baby universe were to form out of a local destabilized space and time in our own universe, so goes the theory, it would decouple immediately from our own units of space and time through a wormhole effect, a phenomenon that was first proposed by Albert Einstein and Nathan Rosen in 1935, and is today referred to as the Einstein-Rosen bridge. Some may argue that we already have circumstantial evidence that these wormholes exist, by the mere existence of our universe itself!

Our own universe would be a decoupling event produced by yet another universe or from universe pairs, so goes the theory (similar to what we discussed before about the forming of negative and positive time universes). In other words, God builds Einstein-Rosen bridges, although I think he knew what they were before Einstein and Rosen figured them out! And yes, the birth of our big bang itself could have been nothing more than the output result of one of these wormholes.

Thing is, wormholes are dangerously unstable, if indeed they can even normally exist in nature. To create one, you would need to produce and maintain an enormous amount of energy (10^{28} electron volts, the Planck energy) in a very well defined spot (10^{-35} meter, better known as the Planck length). If you can't do that, you might want to look around in nature for one. Not a bad idea, but even then you have to find stability; the wormhole might be happy to take you, but it may not be quite so happy to bring you back or even get you to where you wanted to go. Wormholes in nature need a point of negative energy density to counteract the attractive positive energy density around it. Negative energy that gives rise to antigravity is one explanation for the expansion of normal space and time in our present universe; and a localized negative energy density would produce a local warping of space and time with opposite curvature in our universe thus allowing one to "shortcut" across normal space and time. Since this localized kind of natural wormhole would just deposit you back into the universe at some other point in space and time, it could provide a solution to the problem of space travel across galaxies, or even time travel, so goes theory! However, the basic problem with a wormhole is how to gain the energy density necessary to create one, how to make it stable, and how to control it, much less do any naturally exist, or where would one find them.

As it turns out, a very advanced civilization billions of years down the road may just have such an interest to cross through a wormhole out of sheer desperation. The only way out, in order to avoid the final deep freeze of our own universe, when all the matter in our ever expanding universe loses so much energy so as not to be cohesive, and very low energy photons fill up the universe, is for an advanced civilization to travel across a wormhole of unstable space and time into another universe! Perhaps you and I will never need to worry about it, but someone might. The thing is, even if the wormhole trip is successful, you would like to know in advance what is in store for you in that new universe you're crossing into! If this idea doesn't sound appealing, or if one is just afraid he or she just might be crossing into nowhere, then a good suggestion might be to take theological considerations into account, such as why we are in this predicament in the first place, of an imperfect, asymmetrical universe which eventually meets its own death, what caused it, and more importantly, what we can do about it. And the devil, since he is also cast out of the heavens, might have his own designs about what he could do to save himself. For instance, you wouldn't probably want to be crossing through the wormhole with him, not that you would likely have much of a choice if faced with the situation! Oh well, with wormholes you could say that the devil is in the details!

So the question then might be asked, how does one regain this symmetry which would represent our reconciliation and restoration with God? Humanity has, after all, always held out hope for its position in life. (The devil, it would seem, has never shown much of an interest in it, if the myths are true.) Considering that the falling from grace into a world full of disease, hunger distrust, malice, greed, fear, and sin may represent God's punishment for a cosmos that is itself eventually doomed to failure and finitude, then the restoration of the symmetry of the universe that was lost during creation could represent the return to paradise, perfection, immortality, and yes, the grand nothingness and emptiness of space and time itself. Referring back to our Gnostic philosopher friend, Valentinus, this fall from grace wasn't really our fault, however; and thus how could there exist possibilities for a return to the favor of God, namely a return to perfect symmetry with the universe. In fairness to Valentinus, though, he never had the chance to study string theory or Heim theory! Thus, there could be several other dimensions out there, at least one being consciousness; and if that is true, then it may not be necessary at all to consider the resymmetrization of space and time, and matter with antimatter. In other dimensions outside of our physical cosmos, the symmetry could well be maintained; and if so, it is with these other dimensions where we might still be able to regain our bridge, reconciliation, and connection to God. Of course, it may well require us all to die first, but that's a minor nuisance! Yet, everyone and everything inside the normal dimensions of space and time would have to be terminated, especially when the table is set for the cosmos to meet its timely demise, by either an overall grand cosmic gravitational collapse, better known as the Big Crunch, or by expanding apart into the Big Chill. The most distant supernovae data taken in 1998, along with the WMAP data in 2003, seem to suggest the latter scenario is more likely to occur. Thus, when the expanded universe finally cools back down toward 0 degrees Kelvin, we will find ourselves back to where we started, in maybe 10 billion years or so, in the quantum vacuum with virtually no ground state energy. So you might want to hang on to those parkas!

As it is, the cosmic microwave background left over from the big bang has already cooled down from a superhot 10^{32} degree Kelvin furnace to just 3 degrees Kelvin. And with time, every other hotspot in the universe (such as our sun) will eventually cool down in the same manner, to a point where life as we know it won't be able to function very well, in fact not at all. So what could turn out as death by fire for the earth, according to the Bible and as the sun loses enough energy to become a red giant, could well turn out to be death by freezing for the rest of the universe, in ages to come. Then, everyone will be headed for the wormholes, or face trouble. Yet all hope is not lost; other dimensions beyond space and time may provide an answer, assuming we ever find the interface. For us to get there, however, we must find some way to reunite theories of General Relativity with quantum mechanics to find the bridge; and for that, there may just be such a plausible answer. One possible answer lies with a theory of quantum gravity, called string theory; and if it's right, a unification of all forces of nature can be described to find the Theory of Everything, using a hyper dimensional construct for the universe where symmetry, and hence reconciliation with God, are still reachable.

Before we explore this promising idea in more detail, we first take a detour back into our own universe, and see if we've missed something; namely, why are we left alone in an asymmetric and imperfect universe to meet our own fate, especially now since we know something about wormholes? Is it because nobody else wanted to be here – and why is that? And what seems so special about the earth, anyway, that life seemingly only exists here? After all, didn't we already settle that argument between Ptolemy and Galileo about the geocentric center of the universe being somewhere other than here? (In fairness to them however, I think they were only worried about our solar system, not the entire universe.) And as for that altercation between the archangel Michael and the devil, is there any indication from we've learned about our universe that it was created for reasons other than myth?

Well, there are more clues that we keep discovering in the universe all the time that can help us answer some of these questions from a physical point of view; and these answers are what we explore next. But in order to answer the theological questions in parallel, we are forever mindful of Giordano Bruno's predicament of the plurality of worlds, if indeed other civilizations are ever to be found. If they do exist, someone may be out there watching us. And if they don't exist, the equally daunting prospect faces us that the devil only finds entertainment on our rock in the heavens; but, of course, having just one world to worry about would allow God, and his archangels, to maintain their focus on us. So we come to the subject of extraterrestrial life, from what we know versus what we don't, and where that takes us on the road to the Theory of Everything, and our understanding of God.

XI. Calling E.T.

The founder of Scientology and popular science fiction writer L. Ron Hubbard (1911-1986) once wrote "The head of the Galactic Confederation (76 planets around larger stars visible from here) (founded 95,000,000 yrs ago, very space opera) solved overpopulation (250 billion or so per planet -- 178 billion on average) by mass implanting. He caused people to be brought to Teegeeack (earth) and put an H Bomb on the principal volcanoes (Incident 2) and then the Pacific area ones were taken in boxes to Hawaii and the Atlantic Area ones to Las Palmas and there packaged. His name was Xenu."

With due regard to Mr. Hubbard, one would like to know where are all these planets that are inhabited, and whatever happened to this repressive leader Xenu anyway. Perhaps scientists at the Search for Extra Terrestrial Intelligence (SETI) Institute would even be interested. As you might have guessed, Scientology doctrine itself includes the belief in extraterrestrial life, as it connects with other parts of the doctrine about reincarnation and evolution. However, many people throughout the ages, not just Scientologists or SETI Institute scientists have been inspired or intrigued by the idea that we on earth might not be alone. Greek philosophers debated it, and educated people throughout the ages believed it. Many philosophies and religious beliefs were built around the concept. By the 18th century, it was almost common thought, even in theological circles, that a "plurality of worlds" existed (probably a derivation in some form from that of Giordano Bruno); namely, that God in his infinitude could not resist the desire to have several inhabitable worlds available to him. And over the past 70 years, entertainers, cartoonists, sci-fi writers, and jokesters have gained great mileage out of those scary and demonic outerspace invaders coming down to destabilize or destroy earth. One only needs to consider the worldwide response to Orson Welles' 1938 radio broadcast of *The War of the Worlds* about invading outerspace intruders, or even more recent movies on the same subject, to understand the interest in the subject.

But there was only one minor problem. Never in the history of the universe was there any real evidence to confirm or dissuade the idea that extraterrestrial life exists, much less life on even our next door neighbor mars. A billion dollars or more was even spent by NASA on the Viking 1 and 2 space crafts during the 1970s just to find out if any life might have existed or still exists on mars (why the fascination with mars is of course another question). Again, up to the present date, nothing has been found. And in 2005, NASA was at it again by sending the Mars Reconnaissance Orbiter to orbit and take high resolution photos of the red planet. Apparently those little green men are good cave dwellers; or maybe they have just managed to mix in well with the general populace on earth, like the uncle with his hidden antennae under his scalp in the 1960s *My Favorite Martian* hit TV series.

Then there was the real case of those "little green men". One summer in 1967, Cambridge graduate student, Jocelyn Bell, was busy helping her thesis advisor, Anthony Hewish, construct a 4 acre radio telescope out on a plane field in England, thinking that this was a good way to find quasars, superdistant, highly luminous, strongly redshifted, and sharp objects in the cosmos, that are thought to get their energy from a centralized massive black hole that releases radiation as it consumes all nearby stars and gravitational energy around it. As she plotted the radio signal chart data, she soon noticed a 1.3 second beat signal coming from some distant (extra-solar but intra-galaxy) point source. No natural cause in space seemed plausible. As more people got involved with this very interesting development, Bell was becoming increasingly agitated. Here she was, working hard and trying to finish her thesis on a new way to detect quasars when, as she put it, "some silly lot of little green men had to choose my aerial and my frequency to communicate with us." On top of that, the British media had taken wind of the findings, and a circus of attention was forming. Later, by observing another part of the sky, Bell noticed that another similar beat signal was detectable, this time at 1.2 second intervals. Well that was a final breathe of relief because she knew at that instant something natural was happening in the sky; and "no silly lot" of space adventurers were out to get her.

What she and her advisor finally figured out and published was that they were detecting pulsars, a kind of neutron star that rotates on its axis, emitting radio waves that appear to switch on and off at a rhythmic beat to any distant observer on earth, because most of its energy is trapped at its poles by an intense magnetic field. The neutron star itself is an incredibly dense remnant star of a much larger former star that could not withstand the force of its own gravity such that an explosion occurs, called a supernova, which causes an ejection of the star's outer gas layers, while the remaining inner core becomes an extremely compact mix of nucleons. When their research was completed, Bell and her advisor Hewish published their new discovery, while the results of the original work, that was to find a new quasar detection technique, only went into their article as an appendix! But one can't blame them; the discovery of pulsars earned Anthony Hewish the 1974 Nobel Prize in physics.

While scientists love the general idea that there may be serious life forms out there in the cosmos just waiting to establish communication with us, the desire to find such life turns into despair when the actual effort to find it turns up with no results. In fact, this love fest to discover extraterrestrial life began to turn into downright bitterness by the end of the 20th century when increasing piles of evidence showed no traces at all of such life; and the efforts to obtain such data were starting to get quite expensive, pushing even the best patience of many a government funding agency. Of course, theologians these days seemingly have no interest in it, or have turned against the idea all together; at least they certainly show no interest in talking about it. For one thing, it would be hard to know what an extraterrestrial's experience would be with its creator, and how consistent it would be with ours. That there would need to be such a huge investment in outreach and witness programs in order to spread the word would probably create a tremendous financial resource nightmare for even the wealthiest of religious organizations!

However, in all seriousness, it does beg the question, why only us? Enrico Fermi (1901-1954), a preeminent Italian born nuclear scientist who emigrated to the United States in 1938 to escape Muscillini, had asked just the same question, "Where is Everybody?" The question of course was not in regard to his accomplishments in nuclear science, but in regard to extraterrestrials. Yes, even Fermi was thinking about it, after having been the first to discover and produce the first sustained nuclear chain reaction in the world, an event that occurred on a squash court on the campus of the University of Chicago in 1942 with graphite and uranium fuel. One wonders what he was really thinking about though. After Fermi had produced this historic reaction, he had a coded message sent to James Conant, one of the leaders of the Manhattan Project (that produced the first atomic bomb in 1945): "The Italian navigator has landed in the new world...The natives were very friendly." The natives he was referring to were likely the residents of Hyde Park surrounding the campus in the heart of Chicago, who remained safe from the chain reaction! However, with his colleagues, the conversation would eventually turn from nuclear fission using uranium to his interest in outerspace intelligence.

Nothing in science (or theology for that matter) has ever been discovered that suggests that it *must be* only us earthlings who are allowed to live in the expanse of the cosmos, the 18th century accepted doctrine of the plurality of worlds notwithstanding. After all, the very particles that make up the atoms in our bodies and everything that we see on our planet can be found in the outer reaches of the cosmos; these particles form the very basis for all the waters referenced in the creation story of Genesis itself, where the sky formed the "dome" that separated the outer "waters" from the "waters" on the earth. No, there clearly was something missing in our understanding of the cosmos. Fermi's point was that if the cosmos was around for so long, and if intelligence did exist and evolve elsewhere, surely there would have been enough time for some group of smart aliens to have mustered enough technology together to send interstellar ships to all the outer reaches of the cosmos; for instance, even 30 million years is a mere drop in the bucket of the history of the universe which would have been enough time for some civilization to have picked itself up off the muck and slime of its home planet and explored across thousands of light years through at least our own galaxy. The problem became known as the Fermi paradox.

In 1972 and 1973, the infamous Pioneer 10 and 11 spacecraft were launched by NASA, which had plaques on board containing messages about the earth in every form of human communication. The spacecraft were destined by design to travel forever beyond our own solar system; and they are the first spacecraft to have ever done so. They were the first spacecraft to get a close-up view of Jupiter and to go through the Asteroid Belt. A welcome message from earth is inscribed on plagues on the craft, along with pictures of the hyperfine splitting spin states of a hydrogen atom (hydrogen being the most abundant element in the universe); and included on the plaques are the pictures of a naked male and female. One can only imagine why NASA decided to do this; but at the time the craft were launched, there were no shortages of angry letters to the Los Angeles Times about obscenities being launched into space! If only NASA knew at the time how important these spacecraft would turn out to be in our understanding of the universe, but for much different reasons than their original intent!

As for the SETI Institute, it was originally set up for the very purpose to do research on life in the universe, specifically extraterrestrial life; data that they hoped to collect could go a long way to help resolve the Fermi paradox. Carl Sagan was one of its principle members, as it was officially founded in 1984. However, as early as 1960, SETI research had been underway to do large antennae signal searches for outerspace intelligence. By 1971, NASA, and even scientists from the Soviet Union were involved; however, one major project that was designed to establish 1000 earth based radio telescopes had to be abandoned due to its expense. Yet in 1982,

NASA was funded to do high resolution microwave surveys of space; and SETI Institute activities were funded by NASA after its startup in 1984. However, in 1993, the U.S. Congress did it to science again, as they cut funding for the SETI program out of their budget, having criticized the program as being nothing but a search for "little green men" (does the rhetoric sound familiar - like monkey see, monkey do). After the particle physics SSC collider fiasco, what else could you have expected from Congress! Since then, the SETI Institute has been supported only by private donations, with about 120 employees, and regularly searches for signals on two million radio channels using the Arecibo radio telescope in Puerto Rico; and from the Allen Telescope Array in Northern California, it sweeps through the central plane of the Milky Way Galaxy looking for signals of intelligence.

SETI scientists have tried to project just how many different extraterrestrial civilizations there really could be. They have asked questions like how many suitable stars there are in a galaxy like ours, how many of those stars have inhabitable planets, how many planets would each star have, how many planets could develop life, how intelligent the life would have to be in order to develop communicable technology, how willing it would want to communicate, and finally what would be its possible evolving lifetime as a civilization. For instance, one calculation has suggested that something close to one thousand radio-transmitting intelligent and communicating civilizations should exist in our galaxy alone! Well, as you can imagine, this kind of calculation will have its share of doubters as well as believers in the astronomy and scientific community, if for no other reason, why still haven't we seen a sign from just one of them yet, even in our own galaxy! Other scientists get a number closer to just one civilization, namely our own!

And there are the speculations about whether advanced civilizations might have traveled, probed, or inhabited other parts of the cosmos themselves. For instance, they could have spawned robotic probes that could be detectable from earth which were sent out to inhabit the galaxy and then replicate themselves. Unfortunately, as Carl Sagan once pointed out, this would have been stupid to do really; because if a civilization did that, the robots could wind up becoming technological monsters that, not only would have no desire to terminate, but might replicate themselves in numbers that could overrun the galaxy and indeed come back to even haunt the home planet of the parent civilization that created them in the first place, like something out of the movie *The Terminator*! In any case, Fermi's original question and paradox are still unanswered, because none of these speculative models of outerspace civilization has been supported by data or research. So far, there still has been no evidence found even remotely associated with extraterrestrial intelligence. So maybe we can ask Fermi's original question in a different way: Why is there nobody?

Well in 2005, Northwestern University and U.C. Berkeley astrophysicists published a result (Ford, E.B., Lystasd, V., Rasio, F.A., *Evidence for Planet-Planet Scattering in Upsilon Andromedae*, Nature, 14 April, 2005) that may well explain why life is actually so hard to find in other star systems; they found that the first extra-solar star system that has been discovered to have multiple planets has many planet-on-planet gravitational interactions which make their respective orbits unstable. While we've recently discovered the existence of 160 or so planetary systems in the last decade or so, we have had a harder time finding planets with stable orbits; in fact, none have been found. Our solar system seems to be very special in that planets in our system have relatively stable, almost circular orbits about the sun; namely, in the language of astrophysics, they have a low eccentricity value. And good thing too. If you have a hard time adjusting to winter and summer here on earth, you would have had a harder time to adjust to the change of seasons on some of the extra-solar planetary systems that have now been found!

It's unclear why ours seems to be the only stable planetary star system, but so far, all the others have been found to have very eccentric or short lived circular orbits, which are unstable due to the violent confrontations of the gravitational pull of other planets in their star system. This gravitational tidal effect from nearby planets creates a gravitational slingshot effect that can, for instance, shoot one planet far out into space, only to return thousands of years later to its star system, or maybe not at all. The system that was studied by the Northwestern and Berkeley research teams was the Upsilon Andromeda system; it has three unstable planetary systems (the first system that has ever been seen to have multiple planets) with many planet-on-planet gravitational interactions, and one further problem. It seems there was a fourth planet in the system that is, well, no longer in the system. It appears the planet was ejected out into space by the others, with no opportunity to return. Life as we know it would have had a very hard time trying to exist in an unstable system like that; and some of these eccentric orbits create paths extremely close about the stars, only to have the planet ejected far outside the start system during another part of the orbit. The air conditioning and heating systems in those systems might not be quite good enough to sustain life during their extreme seasons! All of which also suggests that future Star Trek voyagers better pick their outpost stations very carefully, or they may be faced with some serious multi-yearly common catastrophes!

Now, before one gets that nice warm and fuzzy feeling that we are the only stable planetary system in the stars discovered so far, one needs to consider two points. One point is that even Uranus and Neptune in our own solar system exhibit this planetary tidal gravitational effect which causes them to have eccentric orbits, albeit to a much lesser degree. The second point is that one should better consider that our next nearest neighbor galaxy is 2 million light years away, the Andromeda Galaxy. It seems that our Milky Way Galaxy and the Andromeda Galaxy have become an orbiting galaxy pair that are themselves on some kind of collision course with each other. In about 3 billion years or so, the two are going to suffer a cataclysmic collision that will create new star systems and eccentric planetary orbits, all over the place. As it is, earth isn't much older than that; but it's a good bet that life on our planet won't survive much longer than that either. Theologically, it may be an interesting point why we seem to be living now, at about the halfway point into the expected lifetime of our planet, not to mention the sun's lifetime, since the sun will become a red giant in about that amount of time itself; then things will really get hot on earth, barring the upcoming side effects from the Andromeda Galaxy. Who knows, maybe the two events will cancel each other out! Whatever it is, God seems to have wanted us to live on earth halfway into our solar system's expected lifetime. It's just as well though; cataclysmic events are not exactly for the weak-hearted!

Then there is the issue about what exactly causes the creation of an ordinary planet in the cosmos, not to mention its probability to be in a star system that could sustain life. Planets either break off from a neighboring star due to an explosion or collision of some kind or are the leftovers of developing nebulae which also causes the star's creation at the center. In the vastness of the galaxies, where stars cohabitate, the probability for star collisions is remote, even over the lifetime of the cosmos. As for stars exploding, it certainly happens; but then, the planets which would be born as a result would simply become comets, because the star which could have given them life breeding light energy would have disappeared! Then there are the planets that form in unison with a star's creation in a developing nebula; and that is what happened with our own solar system. The star-planet dual creation scenario may be as probable as the stars in existence; however one has to remember that the probability for the planet to form at just the right distance from the star in order to receive the right amount of life breeding energy is also low. So while there may be planets out there as vast as the stars themselves, one has to keep in mind, not only the distance the planet must be separated from the star, but its capability to form and maintain a stable orbit.

So now we can have some fun and do our own calculation of the probability of a life-breeding planet to exist in the cosmos. Keeping in mind there are no professional or funding motivations here beyond the curious, we can do a sample calculation by using our own solar system as a reference, in order to find the probability of there existing a planet with a civilization of life as we know it. In order to sustain life in our solar system, a planet would have to maintain a stable orbital radius (low eccentricity) over a possible sweet spot range (not to far, not to close) of 100 million miles out of a possible 100 billion miles that exist within the sun's gravity well (using the outermost known comets for reference that have orbits about the sun); that works out to be a 0.1% probability to occur. About 5 planets out of the 160 that are known to exist in the universe have stable orbits (all in our solar system), which leads to a 3% probability to occur. In a galaxy like our Milky Way, we have 200 billion stars; thus one would take the quadrature sum of all the probabilities of those stars to have life breeding planets, in order to arrive at the most probable number of life capable planets in our galaxy. The calculation would thus be the probability for each star to have such a planet (.001 * .03) times the square root of the total number of stars (200 billion)$^{1/2}$, or about 13 possible planets that could sustain life in our galaxy, given the constraints discussed so far.

When one further considers each star's probability for being the reasonable sized star at a stage of its lifetime when it is capable of producing stable energy needed by a planet to have and sustain life, the number 13 easily drops by another factor of 10. Hence, we are left with 1 or 2 planets that could sustain life in a galaxy like the Milky Way. And our solar system already has three such planets, earth, mars, and venus that were part of this calculation, and in fact were needed to keep the probability for having life breeding planets in our galaxy from being perfectly 0. Hence, it would be statistically unlikely for another planet in our own galaxy to have life. However, if we can assume that our solar system is not that unusual in the cosmos, if astronomers can ever find other stable star systems like ours, then we could conclude that at least one other planet could bear life as we know it in each galaxy that is like ours. Thus, there could be other life out there, but it would reside on very isolated planets (one per galaxy actually) scattered throughout the vast cosmos and away from each other. Since these galaxies are spread apart from each other by millions to billions of light years, the communication systems between these planets would seem a rather daunting task. As for our Pioneer spacecraft, we lost our last contact with them back in 2002; but it's going to take a few cosmic eons before someone else ever makes new contact!

Of course, in order to find new species anyway, we need to consider the possible array of various species that could exist, their eventual probability for acquiring intelligence, their capability to communicate, or even their desire to do so. If a civilization is composed of only dogs and cats over the course of its existence, then it is doubtful that we're going to find them (or they find us), especially when they would be wagging their tails from maybe two billion light years away. And like dogs and cats, we have certainly developed the capability, and very real probability, to annihilate ourselves before ever getting the first message across the void. Of course, there is the matter of what happens if we were to meet up with a species like the Borg or a warlike Klingon race! Perhaps this sounds funny; but it was actually discussed as an issue before the decision was finally made by scientists to launch the Pioneer spacecraft! So now you can guess why all those kind messages on board are designed to be so friendly!

As for the physical communication boundary conditions between life breeding planets, it takes millions to billions of light years to reach any of these places. For a life bearing planet in our "next door" Andromeda Galaxy alone, it would take at least 2 million years for a telegram, fax, or radio signal to get there. And conversations between these places are not likely to last long! In fact, you're going to be long dead before you ever get back a response; the earth and the sun themselves may not be still around when you do! And one hopes the vacuum of space removes the attenuation of the signal. Now one possibility that exists for establishing extraterrestrial communication would be to make use of Einstein's theory of special relativity and travel in a near light speed spaceship, such that the effect of space becomes dilated enough to travel rather quickly to some of these places. But the amount of energy required to reach these speeds in any ship which has mass goes up by the same factor as the dilation effect on the space you are traveling in; so chances are that it will take lots of fuel and not be cheap. Let's take an "easy" example. Let's say you are willing to spend a year to go somewhere quite nearby in our cosmos, say 1000 light years away. Your dilation factor would need to be 1000. If your spaceship costs even a cheap $1 million to travel at a normal NASA spaceship speed of 40000 kilometers per hour, your dilation factor would have changed very little from one because the relativistic effects are negligible at that speed. In order to get to a dilation factor of 1000, you would need to increase your speed to almost the speed of light, which would require a speed that is at least 25,000 times higher than what you would have traveled at during your $1 million trip. If we assumed fuel costs only scaled linearly with speed, then your trip would easily cost $25 billion to reach a dilation factor of 1000, and hence make your trip in one year's time. Oh yes, there is one other problem you might have forgotten: by the time you get back from your trip, your friends and family would have aged 1000 years during your one year of space travel; one hopes you would have given them a nice goodbye kiss before you left!

So the difficulties are enormous and Star Trekkers have not taught us yet how they can reach warp speed; in their terminology, warp speed is that which is greater than light speed. According to Einstein, only imaginary objects will reach "warp" speed in any case; physicists do have a name for such objects – tachyons. Only problem is that tachyons are not people; and they are not measurable or even "seen" in nature, because they do not interact as real particles that we can see. And tachyons, by definition, travel faster than light speed, thus making it rather hard to catch up with one. Too bad though, greater than light speed travel would have made time travel possible!

That there is one other possibility for intergalactic travel during a human lifetime, or even time travel, is hardly worth exploring, except in the case of dire consequences of a very far future civilization desperate to escape the horrors of our own universe's eventual expansive cool down and break up into stardust. That possibility involves the Einstein-Rosen wormhole, allowed by General Relativity and discussed earlier, that uses the laws of physics to allow instant cross cosmos travel or time travel where a proposed instability or discontinuity in the fabric of spacetime would exist. When Einstein first proposed it, he pondered the possibility of a spacecraft that could use it to travel off to some place, and then return to where it started, having come back before it took off in the first place! You understand that I'm not making this up! Thus, the possibility could also exist to travel across the cosmos in an instant. And physicists who ponder this possibility are not exactly just going *Star Trek* crazy. But before you get your Vulcan ears on or prepare your walkie-talkies for those "beam me up Scotty" commands, you need to know something. That something involves the instability and uncontrollability of the wormhole effect. When entering one, you have to pass from a point of positive energy density into a point of negative energy density such that you probably would be annihilated just as soon as you tried it. It doesn't help anything that we haven't seen one in nature either, not that you are particularly licking your chops to see one; it most likely would be an unpleasant experience for all concerned. But the reason that it seems possible is that quantum mechanics makes it so; areas of positive energy, matter, space, and time can meet localized temporary areas of negative energy, matter, space, and time if the area around them positively compensates for the negative area such that the entire total positive density of the area remains the same. It is really the same argument that is essentially used for explaining how our own universe escaped from its own space and time singularity trap into the big bang that started everything; it is also this same argument that was used earlier to explain why there could be a negative universe somewhere to complete the space/time axis along the negative timeline. Only, in the case of wormholes, we are talking about very localized occurrences in our own "positive" convention universe. For the proponents of natural wormholes though, the question begs to be asked, why we don't see new big bangs occurring all the time, forming baby universes which decouple from our own!

Okay then, let's say we actually found a wormhole, or created one in man-made fashion with the enormous energy density that would be required (10^{28} electron volts over 10^{-35} meter). And let's say we are even desperate enough to go through it, or found some way to make it stable. Problem solved then? Intergalactic travel here we come, with Caption Kirk to the rescue? Well, not quite. A couple of things could happen. First, you could find yourself at the epoch of a decoupling new universe; in this case you might just find yourself alone there with God. But you'd more likely find yourself inside a newly forming hot big bang and end up in God's lap in another form that you probably didn't have in mind! Of course, the second possibility is that the stability becomes lost at some point during your travel, whether it be from the opposite end of the wormhole or from some point in the middle that suddenly gets too small for your round frame, in which case you would be annihilated by the positive energy density surrounding your negative density wormhole environment. But finally, the third possibility is that you could find yourself deposited by a fully stable wormhole into another corner of our space/time universe, the most desired option. In this case, a lot of freaky things could happen, because it would mean that the fabric of space/time is not uniform (so far the WMAP data of the primordial cosmic microwave background taken in 2003 suggests it is very uniform across the cosmos). First, you could probably find yourself somewhere where you'd rather not be. Secondly, you could end up in another time period of the universe; forward or backward, it's not exactly an inviting possibility, given the long periods of unlivable and harsh conditions during the natural lifecycle of our universe. Finally, there would be that distinct nonzero possibility that you could find yourself somewhere and at some time desirable. The only real possibility that allows for this desirable occurrence would be if you could travel through a controllable manmade wormhole that presumably an advanced civilization might be able to build; but again, given the energy density involved, you could likely find yourself in a new universe somewhere. And in order to use a naturally formed wormhole in our universe, nonuniformity in the fabric of space and time would have to be found, perhaps locally around maybe a black hole or neutron star; but then it would not be manmade, and thus not controllable.

Remember though that General Relativity predicts that any object of mass will cause a warping in the fabric of space and time about its gravity well, and massive stars have large enough gravitational wells for the warping to be observed, as Arthur Eddington showed with light in 1919. Time actually slows down as you go deeper into the gravity well, and has been experimentally proven! Hence it would seem possible that a manmade wormhole is a possibility, if an advanced civilization could produce the negative energy density (through some quantum mechanical process that effectively produces anti-gravity) required to create the gateway through some localized warped fabric of space and time; but of course there is still the issue about how to control it so that you could go where and when you please. And there are more problems, as you might have suspected. For one, no such material or energy has ever been obtained that could create this negative energy density on a broad or stable scale, although it has been observed with the Casimir effect (see below) on a miniature scale. Secondly, once you had achieved negative energy density, you would have achieved anti-gravity forces along with it. But you might want to be careful about the gradient those anti-gravity forces might exert on your body as you travel through a wormhole; you might just find yourself literally being torn apart or squeezed like a pancake, depending on the direction of the force exertion upon you. It's too bad though; just think how much money you could make betting on the race horses if you could just hop through a wormhole on a time traveling spaceship after the results of the race were known, and then return in time to place your bet for the start of the race!

As it turns out, the universe horizon would provide a natural example of the predicament that wormhole space and time travel could create, like the one that Einstein was worried about, because of the chronology problems that would occur with the Einstein-Rosen bridges discussed above. The horizon actually contains a snapshot of time 0 during the big bang, because it must move with the original symmetrical one-fits-all-force (before gravity decoupled from the still otherwise unified electronuclear force), where time would stand still according to General Relativity. Therefore, it carries with it the imprint and properties of the origins of the universe. Unfortunately, the only way to reach it would be to encounter a probable anti-gravity field barrier (which would be related to that which is theorized to have caused inflation in our early universe) that keeps it on its expanding and accelerating track away from us; but if we could reach it, we would actually find ourselves going back in time during the trip, indeed even to the big bang of our universe itself! And if things seem weird now, they're about to get a whole lot weirder! You see, at the horizon, time has not budged, and beyond this horizon perhaps exists negative time, and whatever universe goes with it that we speculated upon earlier. But more interesting is the fact that the horizon of the universe would be traveling at exactly the speed of light as it existed at time = 0; and being that it stays at time 0, it has maintained this speed without any natural force symmetry breaking having yet occurred. One way to think about this is to realize what happens when a spaceship travels at the speed of light or what happens when one falls into the center of a black hole. Your acceleration causes your clock to slow down, or indeed stop if you have reached light speed, as directly required by Einstein's equations. So then, if the horizon were going at the speed of light, as it was at the beginning of time, any change in light speed as we go forward in time would be measurable with respect to the horizon. Too bad we can't see it! And getting there wouldn't be easy, as we mentioned. But Einstein's equations would allow many fundamental constants of nature to change, because the fabric of spacetime could get incredibly warped at the horizon. For one thing, a reference frame near the horizon could not be at rest with respect to gravity, so that light speed would not be guaranteed a constant value in General Relativity, for instance, because the horizon would exist near a very rapidly changing gravitational force field caused by the geometry of the spacetime horizon, if not anti-gravity itself that is causing the universe's expansion in the first place.

So what could be going on here? Well, for one thing, the speed of light could have been as high as infinity at the universe horizon during the big bang! In a theoretical publication that we'll discuss later with the issue of light speed, the authors actually discuss such a possibility; but they consider it as a phase transition in time as being a catalyst for a greater than normal speed of light, before it would have subsided to its present constant value. Many theorists today have actually started to ponder whether light did in fact have an early cosmic age speed that was much higher; and this of course is one of the things that could have happened in our early universe. A value that keeps popping up in fact is 10^{10} above today's constant speed of light value, in accordance with our models that use this interaction rate for the gravitational force, and thus the unified symmetric force of the universe during creation. Hence, we have unstable wormholes, where localized warping occurs in the continuum of space and time, and the universe horizon geometry, where dramatic gravitational fluxes could create localized enormous changes in the speed of light, which offer the only two hopes for cross galaxy travel and communication with extraterrestrial intelligence, and neither of which has been observed.

Another thing to consider is that who ever said that the universe only extends out to 10^{26} meters? You would have thought so if you simply took today's constant speed of light value times the age of the universe. But there is circumstantial evidence that the universe is very much larger than that! The lack of magnetic monopoles would be one thing; it was this very problem that led Alan Guth to the theory of inflation in the early universe in the first place. Another piece of evidence would be the extreme uniformity of the cosmic background radiation (density and temperature) as the data shows from the 2003 WMAP results. The universe appears flat with no preference in direction, even in the very early universe, out to the edges of the visible spectrum. But the key word here is visible; there is no reason to believe it stays this way out to the edges of the universe, beyond what can be seen. After all, how did so vast and uniform an area of space that we can see get to where it is, so early in the age of the universe? Models of inflation have the expanse of the universe going out to 10^{25} meters during the first moment in creation; and linear models of expansion that we used for the gravitational interaction rate (with a unified force interaction rate of 10^{18} meters/second, which is 10^{10} above today's speed of light) give an expanse of space out to 10^{36} meters. To give an idea of the comparison of this latter total space versus that which is visible, if something were fast enough to travel out to 10^{36} meters over the course of the age of the universe, the time it would need to travel the visible universe back to earth (10^{26} meters) would only be one year! Thus there would be certainly plenty of space and time in that region outside our visible universe for the fabric of spacetime to do some very weird and funky things!

But now we come to the fun part in trying to explain the significance of these issues for our universe; for one thing, we haven't yet gotten around to solving the chronology problem with Einstein-Rosen bridges. Certainly, riding around on something going faster than the speed of light is sure to cause problems for chronology. Einstein stopped worrying about it because he realized that you would have to become an imaginary entity in order to do it. However, at the same time, he realized an anti-gravity field produced by a wormhole would allow chronology issues to occur. Thus, it would be reasonable to assume that something in nature must keep violations of casualty from occurring. So what might nature be telling us?

For one thing, the 2003 WMAP cosmic ray background radiation data was mapped in all directions and showed no evidence of non-uniformity of spacetime, so that if wormholes exist, they must be quite localized in nature. As for the universe horizon and its anti-gravity field, they would constantly be in motion; they have qualities like a wormhole in the sense that they would allow travel back to the beginning of time. But one must realize that you could only get there as fast as today's speed of light c until you actually reached the anti-gravity field, where your speed could in principle increase until you reached the horizon speed at time = 0, where it could be as high as infinity (keeping in mind that you would still never be able to go above the *local* speed of light in the area)! However, when you would return, you would reverse the process by going forward in time; and your spaceship would start out at this high speed and quickly descend to our known speed of light. But this whole process involves time negating time for you going back and forth, so that no opportunity for casualty violation would occur; and your clock on earth would have never stopped ticking forward by the time you got back.

Now that doesn't mean freaky things could not happen if you could somehow ride a wave of anti-gravity in order to travel back to the universe horizon. For instance, if you got there and took your father with you, cosmic time would still be 0; it would seem that both of you would be in existence before the earth, moon, stars, or any of humanity were born, much less you or your father. If your father died on the trip, for instance, you could come back without him, even though it seems like he died before he, you, or any of us was born! But this doesn't violate a seemingly obvious case of casualty, because all clocks in the rest of space would have still been going forward, except the horizon clock, which simply wasn't moving. Also, you could only come back to some time in *the future* after you had originally left because of the cancellation in time required for traveling back and forth. And in fact, the earliest time you would arrive back would be the time of original departure + delta-time, where delta-time would be the amount of time you spent on the horizon's edge. Thus, you would still return back to the same time that you left, at a minimum.

The same thing would occur with race-horse betting. Now, if the race track had an off-track betting facility (many do these days), but on a different planet near the edge of the universe horizon where time travel was necessary to reach it, you could certainly wait for the race results here on earth to be known and then race back to the place where time was going slower at the off-track facility, and be there even before the race occurred on earth, at least according to the local clock on the wall. But the gig would soon be up, because by the time you got there, the news of the race would have traveled with you. And if you tried to beat it by reaching the universe horizon itself (back to time 0) and then return to the distant betting facility to quickly place your bet on the result you already knew about, it still wouldn't work, because the time it took for you to travel back and forth between off-track betting facility and the universe horizon would negate each other, and you still could only arrive at the facility at a minimum time that the results of the race on earth were just being received! So too bad; but once again, the *problem* could be resolved with a man-made wormhole (by using anti-gravity that was not natural to the area in question so that you could beat the local speed of light!). Then you could make the trip from earth back to the facility (or universe horizon), if you didn't first get ejected into another universe, faster than normal light speed and reach the facility long before the results were known there, with results you already had in your hand! So the final point to all this nonsense is that God certainly did not mean or allow in nature for man to do this kind of thing with the universe horizon or with localized natural wormholes that don't pry open the observed continuity of spacetime.

Yet man, in all his derelict greed, ego, desperation, and desire to place that bet or escape a dying part of the universe, could possibly try to thwart chronology with this exotic man-made wormhole that simply used anti-gravity to find an unnoticed gap in the continuity of spacetime, without an energy density that could divert you to a new universe instead! In this case, an Einstein-Rosen bridge would be formed that would allow you to go faster than the speed of light to a different point in spacetime, where chronology could be shifted to put the future on a tangent from what you were use to expecting. Yet, once again, we run into the predicament of what happens to you, as an object of mass going through lightspeed (in order to change chronology for a race track bet, for instance) and becoming imaginary! God would still end up having the final laugh anyway, if you could call it that, because the perils involved with making such a trip would likely give you a one-way ticket to some permanent place you would rather not have preferred to go! As for a simple fundamental particle however, the problem becomes much more complex.

So we start with the fuel necessary for time travel, anti-gravity, which can only be produced in tandem with negative energy. A good starting point for understanding it would be the Casimir effect, which is understood and has been recreated in the laboratory; it is the only known case of negative energy that has ever been observed. It was first proposed by Dutch physicist Hendrik Casimir in 1948 while working for Philips Research Labs, but only actually demonstrated and measured as late as 1997. The idea was that if you took two noncharged conducting plates and placed them very near each other (but not touching) in a closed vacuum, you would actually see an attractive force between the two. The virtual quanta that give rise to the electromagnetic field behave in such a way that only virtual photons of integer whole wavelengths matching the distance between the plates would exist in this in-between region, thus creating a smaller energy density than that which surrounds the plates, thereby creating an attraction between them. The thing is, the vacuum of this environment implies that only zero energy density exists around the plates to start with; that means between the plates, a negative energy density has to exist in order for the plates to attract each other! In other words, between the plates, something less than nothing actually is occurring, a negative energy density if you will. As it turns out, French sailors in the 1800s noticed a similar effect, but in a different way. It seems they often had a problem; when two ships were harboring near each other in violent waters, a destructive interference of the sea would occur between the ships such that the water between them was actually calm. Yet the ships would be seen to move toward each other, thus creating a natural collision! We know about this today because manuals that warned about the problem and how to handle the phenomenon have been preserved.

So not only do the physical laws of nature not discount the possibility of transgalaxy shortcuts or time travel for particles, they may even support them. But remember that quantum mechanics is driven by probabilities; and particles have higher probabilities of doing things than coordinated substances (like you or me) do. It's the same argument that was discussed about how it is actually possible to be transported through a concrete wall if we stood there long enough; the probability may be nonzero, but you might be standing there for a long time! Thus, these negligible circumstances which are required for time travel finally allowed Einstein to go to sleep at night. But of course, there is the effusive question of Stephen Hawking – where are all the time traveling tourists, if indeed it is possible to do? He tried to answer his own question with his Chronological Protection Conjecture, which basically tries to explain away a possible paradox – if we could travel in time such that we returned 80 years into the past in order to kill that grandfather we hated, then what would have become of us when we returned back to the future, since our father or mother would not have been born? Believe it or not, this question does baffle some of the best minds in physics, since the physical nonzero possibility that anti-gravity wormholes could exist is becoming accepted theory. Even the revisited popular Richard Feynman had concerns about this realistic possibility. One argument is that quantum mechanics makes it impossible for a particle, once it goes one way, to make it back again unscathed. Another argument is that there may be rules of causality that require events in the past to consider every possible path and then necessarily pick the most likely route to the future, which thus must be our own. And then there is the matter whether Einstein-Rosen bridges would apply to other hyper dimensions of the universe, and not just those dimensions of universal space and time; in this case, they would possibly relate to the realm of God and how he uses them to paddle across the universe. Whatever the possibilities for their use, wouldn't it be nice to pose as yourself and go back to the past to tell everyone how it is going to be, or in my case, go to the 1956 Rochester conference to give a talk on parity violation such that I might win a Nobel Prize three years before I was born!

Beyond all this amusement however, you really do get the feeling that someone is trying to tell us something, and it isn't the extraterrestrial beings. If God doesn't want us to communicate with other beings or travel to the outer cosmos, he seems to be having good success putting up the barriers. One would have to relent to philosophers for why this is so, because science simply presents us with the hard facts, and theology skirts the issue altogether. After all, the church is still recoiling from the Galileo catastrophe, not to mention what it did to poor Giordano Bruno, especially since he dared to think about consequences of extraterrestrial life; no wonder that theologians don't like to talk about it very much. Perhaps God just finds this whole idea of human intergalactic travel a waste of time, excuse the pun. And perhaps there really is no virtue in making contact with other potential civilizations, if indeed they exist, beyond the purely curious. We have found however, that if they do exist, they are not in great numbers and would mostly be spread out hopelessly across the heavens. Hence it would seem, for whatever his reasons, God wishes to treat us separate and apart; maybe there really is something to this 16th century argument that earth is meant to be a focal point of the cosmos, even if it's not at the center of things (or anything for that matter). If this were the case, gravitational forces in the universe would likely have something to do with it, as one bizarre observation will show us later, although not in the way they thought about it in the 16th century. As it is, people of faith believe that earth is a focal point for God, anyway. And even that holy and chosen city of Jerusalem itself happens to be at the geographical center of the old world (before the discovery of the Americas), with Europe to the north and west, Africa to the south, and the Arabian and Asian kingdoms to the east. Thus, God does seem to show that he has an interest at the center of things, the center of course being the fundamental feature of all things with symmetry.

One thing we do know for sure, however, is that if there are other life forms out there in the cosmos besides ourselves, they are made up of the same stardust as we are. All elements in the universe are also found as elements on the Periodic Table. Those elements all have molecules made up of atoms that are made up of nuclei of protons and neutrons that are themselves glued together by a set of quarks surrounded by a cloud of electrons. Those quarks and electrons are common in all the elements and particles of the cosmos that make up anything which has mass in the universe, including you and me and everything we eat, drink, smell, touch, and see. The only difference is in the arrangement of those elements, and thus how those beings might appear to us!

We also know that other beings in our universe obey the same laws of physics as we do. And this is implicit in Einstein's laws of relativity. What that means of course is that if we are having trouble communicating with them because of the great distances between us, then they certainly are having the same trouble trying to communicate with us. Even in **Star Trek**, the Klingons and Rhomulins could only go up to the same warp speed that the Federation could. However, the Borg may have been on to something by using their common consciousness to communicate. While this leads us back to our concept of common consciousness, and its window to the mind of God, the Borg after all are just a fiendish species in a TV series! However, what is compelling for us in our universe is that God does seem to give us an underutilized tool of common consciousness whose construction may be able to give us more clues about the cosmos itself, if one were to accept the speculation that this mind of God leads us into a hyper dimensional realm over and above space and time, apart from our petty little universe. To understand these ideas, we turn to the one big hope among mainline physicists today for finding the Theory of Everything: string theory, given its use of the concept of hyper dimensionality, analogous to multidimensional Heim Theory, which has already shown success in predicting masses of fundamental particles in our universe and itself later extended to describe the realm of consciousness.   And what these hyper dimensional theories are telling us is that there may be a road back to the ultimate symmetry of the universe after all.

XII. The Universe Hanging by a String

One day in 1919, Albert Einstein received a letter from a Prussian mathematician named Theodor Kaluza (1885-1954) in which he suggested the possible existence of a fourth spatial dimension. He wrote that if this were the case, the forces of gravity and electromagnetism could be described in the same way. Einstein contemplated this idea for a long time because he was trying to find a unified Theory of Everything himself; but he was deeply troubled how to reconcile his pet General Relativity Theory, which had explained the grand cosmos so well, with quantum mechanics, which had explained the minute cosmos of quantum particles in electromagnetism so well; yet both theories were not in harmony with each other. The problem was that quantum mechanics maps out a set of probabilities for things to occur or be measured in any given place or time, as bounded by the Heisenberg Uncertainty Principle, whereas General Relativity lays out a series of equations that produce definitive results for a quantity's value in space and time. In 1921, Einstein finally replied to Kaluza with a rather innocent observation: if there is a fourth spatial dimension, how come we don't see it?

One can appreciate Einstein's apprehension, if you can call it that. However, in 1926, a Swedish mathematician Oskar Klein (1894-1977) came along with an equally innocent proposal to answer Einstein's question: well if we don't see it, it might simply be because it's too small to see! In fact, he figured out it was 20 orders of magnitude smaller than the atomic nucleus itself! The idea was that the fourth spatial dimension was in size about 10^{-35} meter, in what scientists now call the Planck length; and every particle in space was not in actuality a pointlike object, but rather a circular string with this Planck length radius. If we treat the four dimensions of space and time with one added dimension of this size, Klein surmised that maybe we could combine the theories of General Relativity and electromagnetism as a basis for finding the Theory of Everything. And so the birth of string theory took its shape; it is one of the brightest hopes for theories of quantum gravity, where gravity is included within the framework of a quantum field description of all forces of nature. Unfortunately for Klein, the magic of his ideas got put aside for nearly 50 years because his relativistic wave equation did not fit the experimental results from even the hydrogen atom; the problem was that he did not take into account the spin of the electron. And that was because the concept of spin would not be formulated until just a year later by Paul Dirac (1902-1984), who also postulated the existence of an anti-particle to the electron, called the positron, just a few years before its discovery in 1932. Now the reason the idea of a fourth spatial dimension is of substantial interest here to the nonphysicists in the crowd is not that it so intrigues theoretical physicists today, but that it leads us straight back to where we started – consciousness, the mind of God, and our speculative hyperdimensionality of the realm of God. And quite coincidentally, some psychological models on the structure of consciousness model those of string theory quite closely.

First of all we discuss the physical significance of this unimaginably small dimension. The Planck length itself is derived by calculating a mass where the quantum mechanical description of a particle has equal weight with its general relativistic description. When a particle has a large enough mass that its quantum mechanical length can be known with a spatial resolution (according to the Heisenberg Uncertainty Principle) that is equal to its Schwarzchild radius (the radius at which even light can't escape the particle's geometry due to its gravity, much like a mini-black hole), then that length is known as the Planck length. Now the Planck energy of a particle is that energy where all forces of nature, including gravity, would still be unified inside the Planck length; but it turns out to be pretty big...to the tune of 10^{28} electron volts! If we had the machinery or tools to allow us to see this in nature, we would be able to peer into something similar to the first moment of creation, when all things in the mini-universe were symmetric, conserved, perfect, and the forces of nature were all still combined as one. Physically, at this point in creation, quantum mechanics would by definition have an equal description of nature with General Relativity; namely, the wave function of a point particle at this point would have been equal to its gravitational inertia. And theologically speaking, nothing would have happened or existed at this point that would have been inconsistent with the perfect realm of God or perfect symmetry of the universe. Thus, the Planck length and Planck energy become quite important concepts when discussing where total symmetry in the universe can still exist in nature. And physicists sometimes refer to the Planck length itself, albeit euphemistically, as "God's unit", because it depends only on constants of the universe (assuming light speed is a constant) and removes human measurement, intervention, and convention from the equation; perhaps the framers of this expression may have never known how "on the mark" they might be!

A few confessional side points should be mentioned here before we continue. String theorists haven't really agreed yet as to exactly how many dimensions we're really talking about, whether it is simply a fourth spatial dimension or in fact 6 extra spatial dimensions (supersymmetric theories use 10 dimensions in total); the issue however is only one of math and not principle, depending on how you model the known spatial dimensions in relationship to the fourth spatial dimension of the string. It's probably more general to say that there are several string theories running around, depending on what fundamental concept the author holds more dear; one reason they can't always agree is that there seems to be a mathematical renormalization problem when including gravity, which tends to cause some term in the equations in some theories to "blow up" (go to infinity), not good if you are trying to find an explainable Theory of Everything.

Now string theory has historically been broken down into two kinds, bosonic string theory in 26 dimensions and supersymmetric string theory in 10 dimensions. Both models have problems, beyond the usual renormalization issues, in that the former only incorporates bosons (particles that carry force fields such as photons), while the latter includes fermions by making the particle string simply a vibration of a boson one moment and a fermion the next. Of course, none of these vibrations has ever been observed, and both theories don't handle the concept of mass very well. Okay, so don't give up, even if you're lost. Keep in mind that string theory is really still under development by theorists, and is discussed here as a general mathematical formulism for modeling the universe with hyper dimensions beyond space and time, in order to find a possible solution for the Theory of Everything, including a solution for the unification of all forces of nature, in the quest for the ultimate symmetry of the universe which is missing in normal space and time.

Most recently, another subclass of string theory has evolved, called M-theory, that adds higher or even infinite dimensions on top of our present space/time dimensions that are, in turn, on top of the Planck size dimension(s) of normal string theory; in this case it becomes possible for multiple universes to live side by side or to combine under the influence of gravity and cause another big bang, or another universe to form. For one thing, in this representation, our physical universe could become a small subset of God's hyperextended realm. The reason we would not see it is because the resolvable demarcations of the higher dimensions would exceed our ability from our subset universe to see outward enough to notice them. From a theological standpoint, the realm of God, which should be perfectly symmetrically designed as consistent with his perfect existence, would have to have an imperfect element, namely our subset universe. One way out of this seeming contradiction (that a perfect God would create something that is imperfect) is to offer the suggestion that whatever was imperfect in one set of dimensions could be counteracted and symmetrized by a reverse universe in another set of dimensions, all of which fit under the hyperumbrella of the total realm of God. Thus, a perfect God could create what seems like an imperfect local universe in order to resolve one issue, such as a place for Lucifer in the biblical epoch battle of the heavens between Michael and Lucifer the devil, which then can in fact be counteracted by another imperfect local universe with a common reference or pivot point, in order to maintain the perfection and symmetry of the whole outer realm of God. If any of this still seems illogical, just remember the biblical warning of the apostles, which states that the foolhardiness of God is greater than the wisdom of man. On this point, I rest my case.

In understanding one representation of M-theory, however, we can consider a universal geometry where the extra dimensions of the universe have a radial length R which becomes equal to that with radius 1/R. As we'll see below, this happens in much the same way that we model consciousness as being able to wrap back on itself (the smallest becomes the biggest frame for instance). The interesting thing about this approach is that if you look out at the projected size of the universe, which is now 13.7 billion years old, you get a radial length of 10^{26} meters, assuming the speed of light has been constant from day one till now. The inverse of this size is well above the Planck scale length that is required for universal symmetry. However, if you were to take the idea that the speed of light from the big bang till now has covered 10^{26} meters and that inflation of the early universe was the result of a primordial gravity wave, for instance, which had an interaction speed (g) that could have been at least 10 orders of magnitude greater than today's light speed c (an assumption we made earlier when modeling gravity's repulsive interaction rate across the timeline barrier during creation; also remember this gravity wave would have a speed frozen in time at the universe horizon), then you get a universe of size 10^{36} meters along one dimension, since light speed is $3*10^8$ meters/second (g would be $3*10^{18}$ meters/second) and the universe is about $(1/3)*10^{18}$ seconds of age, give or take an eon or two. Then the inverse of the horizon becomes the same order of magnitude as the Planck scale length!

Thus, starting from string theory, we can speculate on a more general theory that uses a primordial gravitational rate, above today's speed of light, to connect the universe horizon expansion rate with a symmetrical dimension of God, the Planck length; however, in order to explore this speculation of such a dependent relationship between the two requires some math. Since we're not trying to write a research paper here (although it's tempting), we would rather not bore the reader with mathematical details. But we can make some general observations, assuming first and foremost that this relationship could be real, thus giving credence to the idea, the bigger things get, the smaller they seem to be! For starters, the Planck length and universe horizon length scales would be dependent on cosmic time, since the universe is continuously expanding (at the assumed gravitational rate g that is frozen at its value at time 0 by virtue of the fact that the universe horizon clock would have never moved from time 0); we are assuming the universe horizon length would grow linearly with cosmic time and in proportion to $c^{3/2}/G^{1/2}$ (where c is the speed of light and G is Newton's universal gravitational constant), because the Planck length is mathematically defined as proportional to the inverse, $G^{1/2}/c^{3/2}$. Secondly, the nonzero dependence of the speed of light c and the gravitational constant G on cosmic time would also have to be considered; for instance, G would actually have to decrease nonlinearly with cosmic time, and thus not be constant, in order for the horizon to continue expanding linearly with cosmic time.

Now the Hubble constant, in General Relativity, is defined linearly as the inverse of cosmic time; it is the proportionality constant that equals the ratio of an observed recessional velocity (redshift) of a cosmic object to its distance from earth; space probes have measured it by mapping the recessional velocity of the stars versus their distance from earth. And its value has been measured to be eerily close to the inverse of the actual cosmic age, as it should be; and that's interesting just by itself because there are issues left unresolved about why this Hubble constant is a measure of cosmic distances that are spatially warped by the linear expansion of the universe, but it can't explain the spatial accelerated expansion of the outer cosmos, an issue that so far has been swept under the rug with the concept of dark energy. For instance, could there be another time component to it? But if the universe were to expand linearly at the primordial gravitation interaction speed (speculated to be 10 orders of magnitude above the speed of light, which we've denoted as g), then it would turn out that the universe horizon distance would not only be proportional to $c^{3/2}/G^{1/2}$ and equal to the inverse of the Planck length, but it would also be equal to g times the cosmic age of the universe, and further, be equal to g divided by the Hubble constant.

In this specific model, one would see then that the revered Planck "constant" would change by one part in 10^{10} every year, just as the universe would increase in size by one part in 10^{10} every year; and hence $G^{1/2}/c^{3/2}$ would change by one part in 10^{10} every year, not very much! Given that we only know G today to one part in 1000 (and c to one part in 10^8), this change from year to year would be quite hard to measure! It can be noted also that in a different model where the universe horizon expands only as today's constant speed of light (instead of g), the Planck length and universe horizon could still be linked; but then the horizon expansion and Planck length would have a different dependence on cosmic time. While these models seem far-fetched, all we really did here was link the Planck length with the size of the universe to come up with a model of dimensional symmetry to explain the geometry of the universe itself. And there is credence to this idea when one considers that the universe horizon is itself a snapshot of time 0 (because time, in relativity theory, would have never budged at the horizon that is traveling at the primordial speed of light or gravity!); and hence the universe horizon, at the seemingly farthest expanse of the cosmos possible, is also a snapshot of the singularity point that was the big bang itself!

One analogy we can make with this proposal is the relationship of mass to size. In quantum mechanics, the effective size of a particle that can be seen by another particle is proportional to the inverse of its mass. Conceptually, that may seem strange: the fatter something is mass-wise, the smaller it appears to be; but remember that we are considering mass and size from a quantum mechanical point of view. Crudely put, the more massive a particle, the better we can locate exactly where it is. It also speaks volumes about what mass actually means, and why physicists should start to pay more attention to Heim's multidimensional theory about how mass comes about; it simply uses the interactions between dimensions that we can see in space versus those we can't in order to bring mass into existence, albeit the theory runs in competition with the Higgs model in particle physics. Certainly, quantum mechanics is screaming out for such a relationship between mass and size. Comparing the relationship of Planck length to Planck energy (10^{28} electron volts) is like comparing the size of the universe to the quantum mechanical mass fluctuation (10^{-41} times the proton mass) at creation that brought the universe to its present age, and which we calculated earlier. In each case, we are comparing a time of perfect symmetry in the universe to our present universe, by simply making use of Heisenberg's Uncertainty Principle, although the only useful observation here in each case is that one is as extremely small as the other is extremely large.

Another interesting facet of this idea is the mass density of the universe. Because density must be continually decreasing as the universe expands (assuming the total mass of the universe is constant), we see that it should actually change as the third power of the ratio of the Hubble constant to the universe horizon speed as seen from earth; that's because the spatial volume of the universe is in three dimensions, and General Relativity shows us that distance in each dimension should equal the universe's recessional speed from earth divided by the Hubble constant. Now since the Hubble constant depends inversely on cosmic time (t) by definition, we have that the density of a linearly expanding universe should then go down as $1/t^3$. As it turns out, the critical density of the universe goes down with cosmic time as the square of the Hubble constant, or hence as $1/t^2$. The critical density is that density above which the universe would resist continued expansion. Since all evidence seems to show that the universe is still expanding, not only are we below that critical density, it would appear that the density of the universe should continue to decrease at a faster rate than the critical density of the universe, thereby guaranteeing the continued breaking apart of the universe in cosmic ages to come. Some numbers can be offered here that are easy to remember, using units of hydrogen atoms, since 80 percent of the known mass of the universe is made up of simply these atoms. The critical density of the universe is equivalent to about 6 hydrogen atoms per cubic meter. Even though the density of the average galaxy is about 1 million hydrogen atoms per cubic meter, the density of the vast space that is in between is only about 1 hydrogen atom per cubic meter (very roughly), thereby explaining why we seem to be underneath the critical density today, and are increasingly likely to stay that way. We'll see that this continuous ongoing reduction in universal mass density may well play a role in understanding the remaining cosmological issues, such as the Pioneer anomaly, fast galaxy rotation rates, and the accelerating expansion of the universe.

A few more observations, however, can be made about the universe's density dependence on time with this model. One sees that as time goes to infinity, the mass of the universe would break up as the universe continues to cool and expand; that doesn't spell good news for your typically advanced, cosmically aged civilization down the road. Also, while the density of the universe would have started off from a singular point of infinity, it would have quickly broken down below critical density, almost instantaneously in fact, thereby guaranteeing universal expansion from the beginning of time. In absolute terms, we can do a calculation to see that the universe is expanding at a rate of one part per 10^{18} every second, or one part in 10^{10} per year, whether one respectively uses the speed of light c or the gravitational interaction rate g for the universe horizon speed. However, if one uses g, one sees that the total universe is expanding by the size of the visible universe (that amount which light could have traveled throughout the age of the cosmos) each and every year, since the universe would be actually 10^{10} greater in size, or 10^{36} meters! And if this were the case, all issues involving flatness and uniformity of the visible horizon could be well understood, inflation theory notwithstanding; and basically all we had to do is assume a primordial gravitational interaction rate g that is 10 orders of magnitude greater than today's speed of light c.

So conceptually, one could model the universe, like models of consciousness, as wrapping back on itself from its smallest dimension (the Planck length) to its largest dimension (the universe horizon), the one actually being modeled as the inverse of the other, and coincidentally be able to find solutions to some basic unresolved problems of cosmology that today can still only be explained with inflationary theory, thus giving merit to the expression, the larger things seem to get, the smaller they seem to be. Hence there would be other implications for this model. For instance, it would imply that the universe is actually closed in some way, giving impetus to the idea that only an organized intelligence could have designed the universe so as to avoid the consequences of the second law of thermodynamics (that all closed systems must become chaotic in time); and it would give a basis for theories about how the vacuum of space itself is responsible for galaxy rotation rates and the Pioneer anomaly that are only explained today by dark matter, which so far hasn't been found. And there would be one other implication of a faster than light speed gravitation rate; if the forces of nature were indeed unified at creation, then the electromagnetic force's mediator photons of light would have equaled that of gravity at the time of the big bang. And that would indeed provide an explanation for the observed accelerated expansion rates in the outer cosmos; an explanation that does not involve dark energy. But before you start to conclude that the speed of light in the early cosmos was so fast as to explain why the universe only seems like it is 13.7 billion years of age, there are other issues to consider, not the least of which are the biblical literalists aching to say I told you so. Later we explore these possible solutions to today's cosmology problems in more detail; but we'll also see that the speed of light, if indeed it was faster in the very early universe, will not explain away the age of the universe. (Meanwhile, a memo to the next generations of Einsteins, Fermis, and Feynmans: if you want to win a Nobel Prize in physics, show why light speed could have been different in the very early universe, provide the relationship as a function of time, propose an experiment to measure it, and last but not least, do the measurement!)

As for string theory, it becomes important because it uses the Planck length to create a string of this size for each quantum particle in order to find a common ground between General Relativity and quantum mechanics and to find a unification mechanism for the time after creation when gravitation was a significant and combined partner of the other universal forces of nature. Unfortunately for this and other quantum gravity theories (one of the reasons there are several of them), there is no good way to experimentally test them. The greatest accelerators in the world cannot resolve matter below 10^{-20} meter (such as when the top quark was found back in 1995 at the Fermilab Tevatron Collider); and we still would have 15 orders of magnitude of resolution to go in order to actually "see" this string that is the size of the Planck length.. Wow, good luck with such an experiment!

Fortunately, we do have some evidence for the stringlike behavior of matter, but it is circumstantial at best. As we study the strong force through the study of the structure of the proton (containing quarks bound by gluons via the strong force), we can model the dynamics of the strong force on these quanta and find that they have a very stringlike behavior in their interaction patterns; but that is about as far as we're going to go in performing tests and observations of string theory. And as we go along, some ideas get a lot of play, if for no other reason, the people who develop them have no fear of ever being proven wrong. And so we get dreams of superstrings and supersymmetric particles and vibration modes of particles that are bosons one moment and fermions the next; the only thing we can say about these ideas is that we have not and will not likely see these phenomena occur, given the undaunting task of probing the size of the dimensions in which they would occur, not to mention the nonexistence of data to support these particle oscillation theories, except for the lovable neutrino which only seems to oscillate with itself. Yet there is this fascination with symmetry in the universe that leads us to find the Theory of Everything; the latter might be obtainable, but the former presents a problem, given the very real asymmetry that actually exists in the normal space and time of our universe.

Similar to the Planck length wrapping around dimensionally into the size of the full universe, one could also model the universe wrapping around into an even higher set of dimensions that we also can't see; this hyper dimensionality of string theory could thus correspond to the outer perfect realm of God, where our imperfect universe remains as a subset. And that is where the idea of parallel universes comes into the picture. For God to maintain perfect symmetry and harmony in his realm, which includes the asymmetric and imperfect harmony of our universe, it would make sense that another universe would exist to compensate for the imperfection of our own universe. String theory, specifically M-theory discussed above, models exactly this kind of scenario. Earlier we discussed the possible disconnect between a negative time universe and positive time universe giving rise to the repulsive gravitational force that may have led to early expansion of our own universe. In M-theory, a similar kind of situation develops in reverse; two existing parent universe "membranes" become attracted by gravity and collide which give rise to a third baby universe. The scenario plays out very rarely in the theory because of the vastness of the hyper dimensional space in which the universes would reside. The parent universes, so goes the theory, would fly apart after the collision; goodness knows where they go!

Now you are probably by now saying, "Give me a break!" Well, yes and no. For one thing, God can choose to do as God chooses to do. Amusingly as a result, theologians probably have more of a problem with these ideas than physicists do, if the life and death of Giordano Bruno was any guide. If we have the right to believe that we exist in our universe, then anyone else in another universe has the right to believe that they exist in a parallel universe. The issue isn't one of believing, but of understanding. Once you take God outside the box, in this case of our space and time universe, all kinds of interesting possibilities exist. And the interesting thing about string theory, for all of its present mathematical difficulties, is that it has created a formulism that can explain thorny theological issues, such as how a perfect and symmetrical God in a perfect and symmetrical realm could produce an environment that isn't so perfect and symmetrical (our universe), but which could interface with his perfect and symmetrical realm, such that total symmetry may be restored under the hyper umbrella of his realm. God could do it through an orthogonal Planck scale dimension; he could do it through a hyper scale dimension that is so big we can't see it; or he could do it by creating parallel universe membranes, where our own asymmetric universe is one membrane that is compensated and counteracted by another. Hence, total supersymmetry of the realm of God would be maintained.

As we have speculated on reasons for the existence of an imperfect asymmetric universe, as it would exist inside a perfect realm of God, we have also speculated on how this symmetry may be preserved within the hyper realm of God, by borrowing concepts from string theory. However, the question becomes more tricky when we ask how one would bridge that gap from our imperfect world to his perfect realm, or try to reintroduce perfect symmetry into our own universe. And that question would lead us to the interface we have, or don't have, with God and his realm. An interesting thing about the Planck length, for instance, is that the secret of symmetry in our own universe lies there, as far as string theory as it relates to the fundamental forces of nature is concerned. If nothing moves outside this length, then all forces of nature can be in unison; and hence the perfection of God's realm would remain. The problem however is that as the forces of nature expanded outside this length during creation, the symmetry of our own universe broke down. At this point, with regard to Valentinian mythology, the universe lost its correspondence with God. Now we can blame the devil for this all we want, but the fact remains we are where we are. So the question is, does the Planck length provide a dimensional interface where our correspondence with God remains intact? Or even more intriguing, does God make use of this interface or dimension for the purpose of communicating, giving power, or providing energy to our universe? However he does it, one thing we know for sure is that God is fast; someone that can make anything go the speed of light, or better yet, 10 orders of magnitude above it, certainly knows how to cook. And from the Planck length to larger dimensions beyond which our universe can see, we know that God can be quite big and quite small all at the same time. So then, it seems reasonable that God would maintain some kind of interface with our existence, especially since we are not going to outrun him, outsize him, or be able to hide from him. For an example of how this interface and connection might be made, we come to perhaps the most sensitive detector that humankind has ever had or used for probing that which we can't see in our own universe, the human consciousness. Next, we discuss this high resolution window, within the context of what we already know about this intricate interface, to the mind of God.

XIII. The Conscious Interface

Models of structured consciousness usually begin with a driver of awareness and self-awareness, where its inputs are sensation and perception, and its outputs are behavior and emotion. Feedback from consciousness itself and experience of the past combine with the matrix input pattern of our sensations and perceptions to drive our consciousness into a state of awareness that then produces a matrix of output behavioral and emotional patterns for our minds to process. If this is all starting to sound like something out of a software design document for a computer program, you are getting warm. Many scientists and psychologists have started to see consciousness in the light of repetitive data processing that can be predicted and repeated, with the brain as the CPU for processing power, memory as the RAM that holds and stores our experiences and past consciousness, and our sensory detectors (eyes, ears, nose, etc.) as input data detectors and collectors to the brain. The key here however lies in the state of awareness, and where it comes from; it is the "I am" in all of us. To be aware of something is to have the ability to understand the information that is being processed from various input states to various output states by the brain. Some will argue that this ability is akin to a separate fundamental force of nature that has a field associated with it which our brains can perceive through a separate sensory detection mechanism, not unlike the force of electromagnetism which is a field that our eyes can pick up as a detector for its mediator photons of light. So far physicists have not taken the concept of modeling consciousness as a fundamental force seriously, much less its possible unification with other forces of nature.

But what physicists have begun to notice is that psychologists model consciousness in a quite coincidentally similar way to the way those extra string theory dimensions of the universe are modeled. String theorists model extra hyper dimension(s) of the universe which are so small, and so large, that we can't see them, such that they are respectively wrapped by and wrapped to the physical dimensions of our universe. Psychologists model consciousness with an endless set of concentric frames which get smaller and smaller as time goes on, such that these frames have a common center and nest inside each other; this is the exact same way that physicists treat the extra hyper dimensions in string theory! The fact that the physical universe can be structured in the same way as consciousness is structured is probably not just an accident or coincidence. If consciousness is indeed a window into the mind of God, then God might have provided himself with the ability to influence, see, and even control our known physical dimensions of space and time through these extra dimensions, which also hold the keys to the symmetry of all the known forces in the universe, and which physicists are loathing to find. In both cases, as you reach the inmost central frame, the whole process wraps back on itself because the innermost frame is identical to the outermost frame, much like what you would see on your TV set when the camera that's producing the picture on your set is focused on nothing but the picture itself. So, you have the string theorists mapping their extra space/time dimensions in this way in order to produce models of grand unification with the fundamental forces of the universe on one hand; and you have psychologists mapping concentric frames of consciousness on top of each other in order to better understand how consciousness is controlled by the brain on the other hand. At the center of the concentric frame of consciousness is the "I am", or Yahweh, in all of us, while at the center of these extra dimensions in string theory is the fundamental quantum unit of nature, the Planck length string. Unfortunately, physicists and psychologists have not really gotten around to agreeing, understanding, or believing what this connection between the two concepts really means, except that they have been developed quite independently and coincidentally to each other. Those who believe that a true relationship between the two concepts really exists have begun to include consciousness itself into string theory; but not many people have followed suit. One person who tried it was Burkhard Heim, by using his analogous multidimensional theory to come up with his own Theory of Everything; but as we found out, many physicists have shunned his work. However, we can speculate that a physical to psychological bridge really does exist between consciousness and the space and time dimensions of our own universe; and common consciousness comes about when the orthogonal dimensions of consciousness act uniformly on all space and time. The only way this could happen would be for all space and time to have a common intersection point somewhere in this orthogonal realm; and that realm could correspond to the symmetric realm of God, thus providing our common consciousness a window to the mind of God, and our imperfect and asymmetric state of existence a channel for communicating with his symmetric and perfect realm. Ultimately, that is what all people of faith strive to do anyway.

Now there are other aspects to this potential connection that we can pursue. Let's suppose that consciousness can be described by string theory, and thus projects out of normal space and time like a set of concentric circles along an orthogonal radial set of dimensions. We all know how we gain energy from feeling love, hate, greed, or fear from people or things that surround us. This energy is provided by our brain to process and instruct our muscles into action or reaction, such as the heart; but this energy is being provided by the conscious part of the brain itself. If this is the interface that God uses to communicate with us, it would not only lend credit to the idea that our consciousness is a window to the mind of God, but that these dimensions along which consciousness resides is able to instruct, command, discipline, change, and tune into all dimensions of space and time in which we reside. Now you can possibly see where this is going. When we talk about miracles occurring on earth, revelations of God on earth, visions of the supernatural, prayers answered (or rejected!), Jesus' resurrection from death, and so forth, one can begin to gain insight into how God does it. As God propagates and projects himself along the coordinates of space and time at each and every point along the axes of our universe, he is able to control or coordinate activity in our universe, as he chooses, from another set of symmetric dimensions from which he has an interface. This dimension remains in perfect symmetry with the fundamental forces of nature in the cosmos, so that if we invoke Valentinus, the gateway to our relationship to God, and thus Paradise, would still exist. Consciousness itself may reside along these dimensions. As for all life with conscious that has died before us within the dimensions of space and time, there is no reason to suggest it wouldn't still reside within these symmetric but orthogonal set of dimensions, such as that which could include our own consciousness, all in harmony with God's realm; only the physical mind interface to our physical asymmetric universe would be missing for those who have died in space and time.

So the question might be how God would actually provide this interface to the physical universe via the consciousness that we all have stored in our brains. In order to explain one possibility, we need to digress to a brief discussion of how the brain involving consciousness communicates and works. First, it should be stated that no organ outside the brain has ever been found that has a direct sensory connection to consciousness. However, the cerebral cortex seems to be responsible for all functional activities (memory, perception, awareness, sensory input, motor response, behavior, emotion) involving consciousness. It is the outermost layer of the brain surrounding the cerebrum, stretching from ear to ear, with 4 lobes, and folds back on itself in layers. The higher mammals have more folding; in humans it is typically 2-4 mm thick. (Perhaps it is smaller for the dumber among us!) However, the critical part of human consciousness lies here, from the superb sensory input capabilities of sensory organs, memory, and perception, to output capabilities coordinating our motor skills of muscles, thought processing, and a fantastically detailed nerve network for reasoning, imaging, behavior, emotion, and communication.

As with the entire brain, information is transferred throughout this region through a nerve network of neurons that are arrayed in large bundles and are organized into pathways or channels. A neuron is a single cell with a specific tree-like array of fiber architecture that receives information via electrochemical stimulation on its fiber edge, called a synapses, from another neuron; and the impulse is transferred to its main cell body down to its tail, called an axon, that in turn generates its own impulse to be sent to the next connecting neuron and so forth. Information from the sensory organs and other organs in the body are brought into the brain by the afferent nerves; and the brain's behavioral, emotional, or mechanical response is sent back out of the brain by the efferent nerves to the muscles throughout the body. And here is where it starts to get interesting, at least from a physicist's point of view. Neuron to neuron communication along the nerve networks is done by electron transport, detected as an electromagnetic field along the nerve pathway so that physical placement changes in phenomenological activity throughout the brain can be measured at any snapshot in time. (This image reminds me about what a physics professor used to say during his lectures when I was in graduate school; he would say the electron does "a song and dance" through the field!) However, changes in phenomenal consciousness that are generated by any new input to the brain, for instance, take place through the physiological nerve processes in the neurons; and these physiological changes result in the continual change in the electron transport capability of the nerve pathways themselves, a different song and dance beat if you will, perhaps like hip-hop back to techno back to rock and so forth.

Now if the electromagnetic field activity, via the electron transport network between neurons in the brain, has control over the communication capabilities of the brain, then we could have a clear key holder to the symmetric dimensions used by string theory for mental penetration into these dimensions. The similarity in the model of consciousness to string theory, in the wrapping around and feedback loop of both, is interesting because of the high resolution of the human brain's electromagnetic nerve cell network for conscious communication, due to the thousand-trillion (10^{15}!) of such nerve cell synapses connections that exist between each neighboring neuron in the brain; and hence the conscious part of the brain could potentially have sensitive penetration and resolution into other minute dimensions we can't see, if indeed they exist. As we would suppose that God is most comfortable in communicating from his realm where perfect symmetry is maintained, God may well have designed our brains to be symmetry detectors into his more perfect universe. As this is all supposition, keep "in mind" (sorry) that no one has ever been able to create anything or anywhere close to the electromagnetic nerve resolution capabilities of the brain. For instance, even dogs have a brain with higher processing power and data throughput than the greatest and most powerful supercomputer ever developed by man!

So now, we can speculate with a model to calculate some interesting results regarding this potential penetration of the human brain to the Planck length dimension, for instance. First we need to make some assumptions. We suppose that string theory is right that the Planck length dimension is a symmetric interface dimension for our universe, that God uses this symmetric dimension (the Planck length which is 10^{-35} meter in size) to be part of his realm for making contact with all physical space and time dimensions of our universe, and that each neuron cell connection involved in conscious communication creates an additive electromagnetic wave state *in phase* with each and every other cell connection, such that a superposition of wave states could be formed. That each cell's electromagnetic field waveguide is in phase with each other is not a bad assumption, considering that it would be coming from a common source, that source being God!

Now we can play with some numbers. W.R. Adey (*Electromagnetic Fields, the modulation of brain tissue functions – A possible paradigm shift in biology*, International Encyclopedia of Neuroscience, Elsevier, 2003) did research to show that conscious controlled physiological and behavioral sensitivities occur in the brain at electric fields down to as low as 10^{-7} volt/centimeter under low frequency and .1 to 1 microtesla magnetic field conditions. That is the equivalent of one ten-millionth of the electric field required to give an election one electron-volt (1 eV) of energy traveling over one centimeter of the field. Now if God were trying to communicate with our brains from his symmetric realm, through the Planck length dimension for instance, then any electronic stimulation seen by the brain from this dimension would be sharply reduced by the ratio of the Planck length (10^{-35} meter) to the normal quantum mechanical size of an electron (10^{-13} meter). That ratio would thus be $10^{-35}/10^{-13}$, or 10^{-22}. However, since there are 10^{15} neuron cell synapses in the brain, a field source from the symmetric realm of God (Planck length dimension) applied to each synapses that is in phase with each other could produce a superimposed field amplitude signal in the brain which is coincidentally $(10^{15})(10^{-22}) = 10^{-7}$, or one ten-millionth, of the normal electronic stimulation one synapses would receive from the same amplitude stimulation in normal space and time. Thus, only 1 electron-volt of electronic energy per centimeter would seem to be enough to create physiological and behavioral communication and stimulation between the brain and the symmetric realm of God with these circumstances.

Two other observations can be made from this analysis. First, superimposed phenomena from all these synapses coming from a hyper dimensional Planck length dimension could give the brain a potential resolution of $(10^{-35})(10^{15})$, or 10^{-20} meter. Secondly, interaction rates in physics experiments are usually measured in units of a two dimensional cross section; so that if we take the square of the brain's resolution of the Planck length, we get a two dimensional measure of the possible interaction rate cross section of our brains to any uniform force field waveguide coming from a supposed Planck length dimension of God's realm; and this cross section would be on the order of picobarns (10^{-40} meter^2). It turns out that this number is close to the smallest interaction rate cross sections that are resolvable today by the world's most powerful particle physics accelerators (such as the top quark's cross section), given the largest center of mass energies (2 trillion electron-volts) that today's accelerators can achieve, excluding the LHC in Geneva. Thus, we can see that the brain could have sensitivity to a symmetric dimension like the Planck length that is near the limit of the world's most powerful accelerators that man can build today!

Well, I'll be the first to admit that these calculations have a science fiction kind of feeling to them with limiting case scenarios with very extreme assumptions. But the real point to be made here is that the brain is a detector to very sensitive physical phenomena that we are not going to find or replicate elsewhere easily; yet our conscious is connected to our own emotions and feelings and our soul itself which makes us what we are. Hence the awareness part of our being, that which says "I am", is at the center of our conscious and would be first to identify God in communication with our universe. After all, not many things in nature like our brains, and hence our consciousness, have one thousand-trillion coordinated sensors for detecting events happening around them in real time! And if the source of communication from another realm or dimension is unique and applied equally to our universe, then the consciousness we all share in common with each other is a channel to that source; namely, common consciousness becomes a pathway to the mind of God. And clearly, God cares about our conscious, if the Christian sacrament of baptism is to be any guide; St. Peter, in his first book of the biblical New Testament, called the reason for baptism "as an appeal to God for a clear conscious". It certainly would seem therefore that God wants our interface into his realm to be a clear one!

Just imagine the electric stimulation that would have been present in the mind of St. Paul as he was accosted by a blinding light and voice on that road to Damascus in the biblical tail of the Christian conversion of Paul, whether you agree with the above model or not. Equally interesting was the lack of sightings by any of Paul's companions; yet they too heard the voice. What we do know in Paul's case, however, is that his visionary senses were overloaded by the light and he was temporarily blinded, as reported in the Bible. How then could Paul have been blinded but nobody else saw anything? It could have been possible if God was interacting specifically with each person's consciousness. At the very least, Paul's sensory feedback nerves in his conscious would have been affected severely by his visionary detection nerve network. And as we have seen, such an effect would also not have taken a lot of electrical stimulation to zap his visionary nerve network. Yet the rest of his synapses worked perfectly enough for him to hear, touch, smell, taste, and record the vision in his conscious memory, a vision that was declared to be from Jesus himself. And of course, Paul not withstanding, in any dream or vision that we have as humans when we're sleeping (or awake!), our visual conscious nerve network is well known to play games of havoc with our stored memories.

It's also interesting to note that animals must also have this communication capability with a God who communicates through consciousness, except that the two dimensional cross section for their detection of some symmetric force within their brains would decrease by the square of the ratio of their brain cells to human brain cells. As arrogant humans, we may be tempted to say that an animal's limited interaction rate with God would therefore deny them access to a spirit, but be careful where you take this line of reasoning. If our ability to interact with God determines our access to our spirit, then mentally challenged individuals or even babies whose brain cells have yet had the chance to develop would have the same problem! Simply because we are human doesn't necessarily give us sole license "to have our cake in heaven and eat it too". We are all part of God's creation, and as such are all inheritors of his creation. The problem here has to do with what we are calculating. These interaction rates with God's realm are a measure of our probability to experience God in our every day lives. His connection with those at a lower level of consciousness in our universe would be indeed at a lower level; but that doesn't mean they don't exist. In fact, it is quite clear that this connection does exist for all of God's conscious creatures by the very fact that this probability must be nonzero; thus a gateway must exist for God to communicate with all his conscious creatures of nature. This gateway just might be useful one day when our imperfect universe ceases to exist, or more likely when we cease to exist in our still existing imperfect universe!

Okay then, the implications of this conscious connection to another realm would be enormous indeed. However, in modern philosophy, consciousness is seen as belonging to a description of one of two possible cases; and not surprisingly, we find that they are a religious one and a scientific one. In the former case, consciousness is somehow taken as independent of the physics of the body and the universe around us, a view that perhaps is consistent with the religious ideal of the soul; in the latter case, consciousness is taken as a direction function of physiology that runs through and continually updates various phenomenological states of our existence.

Obviously, modern science these days accepts the latter view over the former because it sees consciousness as a highly complex, organized information processing, and data thru put system; and we only need to understand consciousness as having two possible representations, the phenomenological and the physiological. For instance, the nerves that control and stimulate consciousness in our brain are simply the result of an physiological electrochemical stimulation of electron transport pathways undergoing continuous phase modulation and propagation delay in order to create resonance output responses to the real-time scenarios that we are constantly bombarded with; yet the brain acts like a phenomenological software state machine that processes real-time thoughts, memory input and output, attentiveness, sensory input and control, and motor output and control. As changes in our ongoing existence occur, the neuron states and synaptic connections are changing in correspondence; they change the capacity and capability of the electronic transport neural network path of the brain, thereby causing a continuously changing, updating, and modulated electromagnetic waveguide to run continuously along the nerve pathways of the brain, dependent at all times on all input data streams and events as they continue to occur in real-time. This highly complex modulated waveguide is not only dependent on time, but it also produces a self-sustaining resonance output response which gives us a sense of continuity, existence, and awareness in the real-time world environment that surrounds us.

The scientific description of consciousness is indeed a mouthful; so let's now try with the religious version of consciousness. There is no question that the scientific view is a very accurate way to see consciousness; but that doesn't mean that the religious view should be thrown out all together. Indeed, the view that consciousness is separate from the body doesn't stand up very well on its own, excuse the pun. However, when taken in context with string theory, the religious viewpoint regains some merit because consciousness is modeled hyper dimensionally on top of the scientific view, which is specific to space and time. And here is where it gets interesting, because this idea would suggest that the brain's conscious has the ability to be an interface, and indeed a symmetry detector, between the space and time and other hyper symmetrical dimensions of the universe. In the religious view of consciousness, the soul is a representation of our connection to God and God's connection with humanity; the soul allows us to connect and communicate between our universe and God's symmetric and perfect realm. The brain, with all of its nerve interconnections and possible electromagnetic resolving ability to a Planck length symmetric dimension of the universe, could be wired specifically to be able to see into another realm; and God, living in perfect symmetry with himself, could reveal himself through this complex nerve gateway in the brain, which in turn sends nerve signals to our muscles and glands that allow us to experience love, compassion, fellowship, generosity, and so on. Consciousness would still be maintained and controlled by the nerve pathways in the cerebral cortex of the brain, so that the scientific viewpoint of consciousness is still very valid; the physiological and phenomenological behaviors occurring within us are still very real. But string theory might thus allow us to enhance our understanding of consciousness by bridging the gap between the two viewpoints of scientists and theologians, not that they themselves are really trying very hard to do so. However, this hypothesis suggests therefore that the brain could be a sensory detector for the religious view of consciousness (e.g. the soul) which exists on top of its scientific description.

From the Scientology view of consciousness, our being is the essence of our soul, called a Thetan, named for the Greek letter "theta". Now, when the soul is named after a letter in the Greek Alphabet, it probably is meant to represent our lack of understanding of it! However, we sometimes refer to God that way (Alpha and Omega) for the same reason. Maybe it is due to some evolutionary philosophical nature that is inside us. Physicists do it all the time!

However, Scientologists believe that the Thetan soul inside each one of us has an analytical mind and a reactive mind that are separate from each other; and the goal for its adherents is to somehow have the former (representing reason and memory) defeat the latter (representing subconscious psychological scars imprinted in us) in order to gain peace in our lives. Well, this is all well and good; but the problem here is that the two minds of the Thetan soul are completely interwoven and linked physically in the nerve processing center of our brains, represented by the ever present modulating electromagnetic field passing along our mental neuron network, a kind of superposition of states if you will, borrowing from physics terminology. Namely, the analytical mind and reactive mind in the Thetan soul are not physically separate and distinct, but rather interwoven and superimposed on our conscious. So now what? Are we forever in a state of despair from our neuroses? Well, let's not go that far. Scientologists certainly can be given credit for giving it "a good try" to find peace. However, there is another way that can solve this problem; and it's a solution we have already addressed. The idea is that God "punches through" the physical space and time dimensions of our universe and accesses our brains through our own consciousness; from the perfect symmetry of his realm, he interfaces with our realm (space and time). The potential success of string theory to explain the unification of all fundamental forces of nature (gravitational, electromagnetic, weak, and strong), in introducing extra symmetrical dimensions which interface with our universe, could possibly provide an explanation for how we can regain our own symmetry and harmony with God, and thereby produce the peace we all yearn for. Our consciousness becomes that sensitive interface to these dimensions; and it allows us to regain our peace with God. When we disconnect from it or ignore it, then we risk the loss of this peace. Someone once said that the brain is a perfectly tuned instrument; if it could be synchronized with the symmetry and harmony of the realm of God, then peace would be assured.

One key to remember is that any new input to the brain will continuously cause it to adjust to a new state of consciousness, while the "I am" part, referring to the center of the updating real-time frames in our soul, stays fixed at the center of the updating frames of consciousness at all times. The updating conscious could be viewed also as the spiritual growth of the individual. The awareness part, which makes us say "I am", is our spiritual center, and becomes reminiscent of God's response to Moses on the mountain, "I am that I am", "I exist", or "Yahweh". And funny I should mention this; but God always seems to find himself at the center of things. The realm of God in our hyper dimensional model wraps around the physical universal dimensions of space and time in perfect symmetry where God would remain at its center; and the center is always the most significant feature of all things perfectly symmetric. As that realm expands outward, then so does our spiritual growth.

The concept of spirit is usually defined as separate from the physical body in most religions, and is usually left to pay the price or reap the reward for the behavior of the conscious body that was/is in our present universe. In Scientology, for instance, the concept of the "Thetan" becomes the "Operating Thetan". In this state, one is able to leave the mind and body and still maintain one's senses, a kind of floating soul, if you will. This gets real close to the model of spirit in other religions, and is of course an interesting subject, although in Scientology the spirit can disconnect from the physical body while it is still alive, whereas in most religions it disconnects after death.

At this point an agnostic scientist would probably begin to snicker, because a discussion of spirit in physical science usually causes one to go off "the deep end". Thus, the best we can do is to suggest that where God exists, the spirit exists. The God we have tried to describe is one who maintains the keys to ultimate symmetry, and any interface we have to him or his realm would have to go through a backdoor, namely an interface, to our dimensions of space and time where this symmetry is maintained. The best that science can provide us today (outside of Heim theory) is that which is dreamed up by string theorists who use hyper dimensions beyond space and time to try to explain what was once the ultimate symmetry of the universe, and the resulting unification of all forces of nature that were implicit in it. Unfortunately, modern day tools, accelerators, and equipment cannot ever hope to resolve these dimensions. The motivation among scientists to find this symmetry is driven by their quest to find the Theory of Everything, because of our inherent human instinct to find perfect symmetry behind everything that exists. For theologians, the motivation is driven by their quest to find God in our world. Funny thing is, the two groups seem to be converging closer than ever before, quite accidentally, reluctantly, and even humorously into a common point of reference. And that point is where they may be surprised one day to find none other than God himself.

Theists often will cite the Second Law of Thermodynamics, which states that in time all systems become disorderly, in order to prove the existence of God, because they ask how is it that the universe is so orderly. When they ask that, scientists quite often respond that a key word is conveniently left out of this interpretation of the law; and that word is closed. The law should state that all *closed* systems become disorderly in time. Hence how can the universe be closed? And in fact it seems to be expanding apart! Well it now turns out that if the string theorists have the right idea that the universe does in fact wrap back on itself from super sized dimensions on top of universal space and time down to miniature dimensions of order the Planck length, the tables would once again, as they say, be turned. Even as the universe expands, there could be a linkage to extraordinary dimensions both small and great that we can't see, which would once again make the universe a closed system. And in this case, maybe, just maybe, the theists would win the argument after all. The funny thing is, regardless who is right about the Second Law, both sides haven't finished the sparring over the First Law of Thermodynamics that got us to this point in the first place: all energy (including heat) is conserved in a *closed* system. The scientists at least maintain consistency by trying to understand where all that heat came from that brought on the big bang; conservation of energy should be maintained, even during creation, especially around a singularity point that surely was a closed system. Theists will suddenly drop the ball on logic when they ignore the physics of this law all together; for them, this initial burst of physical energy simply came from a spiritual God. While there is nothing wrong with this idea for people of faith, we should be careful about choosing which physical laws to quote when others are conveniently ignored. One of the beauties of theoretical science is the requirement that all natural laws and observations must be adhered to, or the gig is off. Theists could learn that lesson. It shouldn't be hard to do though, because I think God is "man" enough to stand up to the heat!

XIV. Faster than the Speed of Light

In ancient times, up to the time of Galileo, light was considered instantaneous; the speed of light (c) was considered infinite. Like today, some scientists had minority views on the subject; but they were largely ignored. Galileo made the first serious recorded attempt to measure the speed of light in 1638 by putting shutters on a couple of lanterns and trying to measure their change in light coverage from a mile away! He didn't succeed. Poor guy, first he allegedly rolled those cannonballs up the leaning tower of Pisa to find a measure of gravity, then he got into a donnybrook with the church over the geocentric nature (or lack thereof) of the earth, and now this. He would die four years later in 1642, still under house arrest by the church. Another astronomer Olaf Roemer (1644-1710) would make a more serious attempt at measuring light speed by measuring the time delay in the eclipses of jupiter's moon io, which he showed to be the transit time of light across the radius of earth's orbit about the sun. He obtained a transit time of 660 seconds while another measurement gave 480 seconds; the actual time is 499 seconds. Unfortunately, his idea for measuring light speed was also debunked and wasn't revisited until after his death in the mid 1700s, when astronomers starting measuring aberration angles of several stars to measure the speed of light; values they obtained ranged from 300,500 kilometers per second during the mid 1700s to 300,000 kilometers per second during the mid 1800s. More precise measurements (down to .01 percent resolution) were made by the time 1900 rolled around, by using rotating mirrors and toothed wheels, where measurements repeatedly gave a speed of light closer to 299,900 kilometers per second. Other electromagnetic techniques were employed by the 1920s to measure the speed of light; those measurements typically gave values around 299,800 kilometers per second.

But a funny thing was happening all the while. A French astronomer named Gheury de Bray observed in 1931 that measurements of the speed of light c were forming a trend; the speed of light seemed to be going down with time. As he put it, how was it that one successive measurement after the other showed the value of c to keep going down. The observation remained very controversial, especially since Special Relativity (a theory that mathematically requires the constancy of light in any inertial frame with no dependence on time) was taking hold with test after test that showed its authenticity. After the war in 1945, most scientists had given up on the idea of a time varying speed of light. By 1956, light speed was measured with an accuracy of 1 part in 10^6, about +/- 100 meters per second, or .0001 percent; yet there was this still annoying tendency for its value to drop, if ever so slightly (1 or 2 standard deviations from measurements taken in the previous century. Finally, by the time the 1970s rolled around, laser techniques were employed that reduced the error of measurement down to +/- 1 meter per second, with a measured value that has remained constant to the present time: 299,792,458 meters per second.

It was observed by Alan Montgomery and Lambert Dolphin (*Is the Velocity of Light Constant in Time?*, Galilean Electrodynamics, Vol. 4, no. 5, Sept/Oct 1993) that if you did a least squares fit to all the data presented for the measurement of the speed of light from 1945 to 1980, however, there seemed to be a statistically significant measurable systematic correlation of light speed with time at a rate of 100 meters per second per year, with 97 percent confidence level. That's a rate of change of .0001 percent change per year! The problem of course was that each datum point as you go back in time has a larger and larger percent error with respect to the rate of change slope; but they claimed that the slope was clearly going down with statistical significance. Also, a nonlinear fit done by Russian scientist, V. S. Troitskii, of the Radiophysical Research Institute in Gorky, U.S.S.R. (**Physical Constants and Evolution of the Universe,** Astrophysics and Space Science, 139, p.389, 1987), where he takes into account measurements of light speed (along with their appropriate errors of measurement) back to 1740, showed that c appeared to have an exponential decay with time, where c at time 0 would have been (believe it or not) 10 orders of magnitude above the present day value of c, or 10^{18} meters per second, exactly what we used for the primordial gravitation interaction rate when modeling our own theories of creation, which would at least imply consistency with the unification of the forces of gravity and electromagnetism before symmetry breaking occurred after creation.

Now the Russian Troitskii could hardly be considered some rebel rousing Bible thumping fundamentalist. But there were even some scientists who managed to fit existing speed of light data into models that showed how light could have traveled 10 billion light years in just a few thousand years; but these models show an even more extraordinary amount of light speed decay than Troitskii had modeled. And these kinds of results are enough to bring out the Bible literalists out in droves to explain how they knew they were right all along. Well, maybe, then again, maybe not. For instance, using even a super light speed of 10^{18} meters per second at creation, it is true that it would only take one year for all the objects we can now see with the Hubble telescope or anything else to actually get here from the most distant reaches of the visible universe horizon. But biblical literalists tell us that the universe began 6000 years ago! (There are some who believe Noah took even dinosaurs onto his ark; I suppose if they didn't sink it, there must have been some wild rocking on board to cause even the strongest of stomachs to weaken!) But with even Troitskii's model of light speed, we would get a speed of light that could not have been more than 10^{12} meters per second 6000 years ago, decreasing exponentially to 10^8 meters per second today; that would mean the most distant visible objects could only be 10^{21} meters away, a far cry from the 10^{26} meters distance from where we see things today by using luminosity analysis. And if we use the least squares model of Montgomery and Dolphin, light decreasing linearly with time at the rate of 100 meters per second per year over a period of 6000 years would give a value that is not appreciably different (less than 1 percent) from the speed of light we measure today; hence those items that are 10 billion light years away must still be 10 billion years old! One thing we can say for the period around 6000 years ago is that the onset of human written language, and perhaps the last evolving mutation of the human brain may have occurred, according to University of Chicago researchers (Chicago Sun-Times, ***Human Brain is still evolving, U. of C. researcher finds***, Sept. 9, 2005); while this is all very interesting, not much of it has anything to do with light speed.

And the physical evidence of the age of the universe just goes from bad to worse for those trying to justify a 6000 year old universe. For instance, the hydrogen and helium content of the universe (80 % and 20 % respectively) versus everything else (less than 1 %) would have taken the universe billions of years to provide that kind of content mix; for instance hydrogen atoms first had to form, and then had to find each other to fuse together to form helium, and so forth, up the periodic table of elements. Then there are the data points from various astronomical objects that show their relative recessional velocity from earth versus their known distance from earth; these data fit a line whose slope gives the present accepted age of the universe as 13.7 billion years, in agreement with General Relativity Theory. And of course, let's not forget that General Relativity has passed every cosmic experiment on light that's been done, except for possibly the outermost reaches of the visible cosmos; and relativity requires a constant value of light speed in any inertial reference frame. That the frame must be an inertial one does provide an interesting opening, however, not to Bible literalists, but to scientists who are trying to solve the other existing cosmology problems today, by modeling light speed as dependent on the fabric of spacetime, but more on that below.

By 1983, the speed of light (c) was set at 299792.4586 +/- .0003 kilometers per second by the National Bureau of Standards; and it has held to that value to within 3 standard deviations since 1973, as accuracy in its measurement in using atomic clock time keeping brought the percentage error down to less than one millionth of one percent! So then international standard committees did something interesting: they chose to define the meter from 1983 onward as the distance light travels in 1/299792458 of a second, thereby freezing the value of the speed of light such that any new measurement change in its value would be seen only in the distance defined for the meter! The issue was left at that. For most approximate calculations that physicists do today, a value of $3*10^8$ meters per second is used as the value of c.

However there were other things still happening. For instance, Thomas Van Flandern of the National Bureau of Standards noticed a slight deviation of the orbital period of the moon that was being measured from 1955 and 1981 by using an atomic clock; and he realized that it was either due to atomic fluctuations used in the measurement of time, or dynamic fluctuations with the moon. Further, the most recent measurements and highest resolution measurements ever made of the fine structure (electromagnetic energy level) splitting of atomic output spectra due to light absorption from source quasars at various red-shifted distances from earth (thus giving a measurement timeline of the quasars throughout the age of the cosmos) have given a possible indication of a time dependent value on the splitting, which may have changed by 1 part in 10^5 over 10 billion years (***Scientists Discover One of the Constants of the Universe Might Not Be Constant***, Science Daily, May 12, 2005), interesting since this fine structure splitting goes as the inverse of the speed of light c; it has been suggested that this could be further evidence of a nonconstant value of c with cosmic time. A very thoughtful paper and detailed analysis by Andreas Albrecht and Joao Magueijo of the Imperial College of London in 1999 (***A time varying speed of light as a solution to cosmological puzzles***, Phys.Rev. D59 1999, 043516), makes a very serious theoretical proposal in favor of a carefully designed time dependent speed of light that could explain cosmological issues such as flatness, horizon, inflation, and the famed cosmological constant (used these days to combine inflation with relativity theory to explain the observed accelerated expanding outer visible universe).

A book by Roger Ellman, (***The Origin and Its Meaning***, 2nd ed., The Origin Foundation, 2004) also makes the argument that when one removes the assumption that certain constants in the universe don't change with time, one can explain away the three cosmological problems that still are unsolved today, such as the Pioneer anomaly, the outer accelerated expansion of the cosmos, and the fast rotation rate of galaxies. (Yes, even cosmology has problems that come in threes!) He makes the case that the speed of light can answer these questions if it is allowed to fundamentally decay exponentially with time, using a decay time constant of 11 billion years. What this means is that light speed would have decreased by about a factor of 3 since the big bang 13.7 billion years ago, in this model. Ellman actually shows that the data from the Pioneer Anomaly, the galaxy rotation rates, and the universal expansion rates all fit perfectly into his universal model, where light has a decaying speed that is slowing down with time. The sideline of this idea, of course, would be that the question of where all that the dark matter and dark energy is goes away, because they wouldn't need to exist to explain anything. In a way, this is like measuring the decay of space and time in our universe. At our present snapshot in time where everything appears constant, the effect would be to produce a higher light speed from distant supernovae in the early universe, and to produce a space contraction factor for even the Pioneers as they speed past our own solar system to a great enough distance so that the resolution of the effect could be seen. Without having to agree or disagree with this hypothesis, one could make the point that the idea of universal decay is quite thought provoking, and fits well with theological concepts of universal finitude and even Valentinian philosophy. As a scientist however, one would want to see experimental evidence and data for the phenomenon (which so far nobody has tried to obtain) before wishing to agree or disagree with the hypothesis; one way to see it, for instance as Ellman suggests, is to study the photoelectric effect (the process by which electrons are knocked out of atoms as a consequence of a certain exact energy of light being absorbed by them) of distant and nearby comparable stars. But once again, the greater scientific community has not really come to any acceptance of these ideas, as evidence (and/or bias) has not been presented to allow it. But gladly, there are a host of other scientists who have begun to infiltrate scientific literature with models that assume there could have been universal changes in constants, such as the article referred to above by Albrecht and Magueijo with regard to the speed of light, in the cosmic distant age of time. Probably the source of some of this renewed interest in the speed of light comes from the fine structure splitting data mentioned above that has recently given possible evidence to a time dependence with the distant past of at least one fundamental constant of nature. Some of these scientists have suggested that any change would have occurred, at the very least, in the very early universe only, which would probably rule out Ellman's proposal; and such ideas are discussed more below.

So what's going on here? First we have scientists who mock the idea of a varying constant like the speed of light, and religiously stick with Einstein's theory of relativity that allows no change in its value with time or other inertial reference frames; they point to the mountains of consistent measurements taken in the atomic clock era that show light speed is constant to within its standard deviation of measurement from year to year. Then we have scientists who believe it's possible for the speed of light to have changed, and point to problems in cosmology that could be explained with a time varying light speed over the ages; they point to data taken over the centuries which show light speed slowly decreasing with time, or recent measurements of a changed fine structure constant taken from spectra of distant cosmically aged quasars. That leaves the rest of us who are not so sure, while stubborn scientists and creationists seemingly having a lock on knowing what they are talking about.

Well there are a lot of issues here. First, when making measurements, you have to know your laboratory equipment, how accurate it is, the systematic errors that might be associated with it, and last but not least what biases come into play in doing the measurement. For instance, with light speed measurements, you have to know two things right away: distance and time. How do you measure both precisely enough for the desired resolution you wish to achieve? For instance, in 1791 the French defined the meter to be one ten-millionth of the length of the meridian going through

Paris from the North Pole to the equator. In 1889, they changed the definition to be the distance between two marks of an alloy of platinum, and measured only at the "melting point of ice". More or less, that definition held until 1960, when the meter was redefined in terms of the wavelength of a krypton-86 source. And later in 1983, as we already mentioned, it was redefined again as the distance light travels in 1/299792458 of a second. Then there is time. How do you know what a second is? As basic a unit as the second is, it too has changed definition several times over the past century! In 1939, the second was defined as 1/84,600 of a mean solar day; it is the mean because the solar day actually changes throughout the year due to irregularities in the rotation of the earth. Thus, in 1960, it was decided that it was more accurate to define the second by taking the measurement after one "tropical" year, instead of taking the mean of many measurements of various solar days. Yet, the irregular rotation of the earth about its axis is also known to be slowing down systematically with time by 1 part in 100,000 of a second every year, due to gravitational forces in our solar system; so in 1967, the second was redefined by atomic means: it is now defined as "the duration of 9192631770 periods of the radiation corresponding to the transition between the two hyperfine levels of the ground state of the cesium 133 atom". But even here the measurement depends on the constancy of quantum energy levels in the cesium atom, which in turn depend on the charge and mass of the electrons and nucleons in the atom.

Then there are the biases that a scientist is trained to reexamine from time to time; we're all human and often wish for a particular result. If we don't get it, we often try again and again to obtain a different one, unless we get the "right" one of course; then we stop trying. And there are the models. One has to be able to fit them to existing data and observations, or else they become void. Also, one has to develop ideas or methods for proving a theory's hypothesis, in this case how to measure light speed over the course of the universe. And that isn't easy, within the confinement of the error resolution required. For instance, in analyzing the unexpected accelerated expansion of the outer universe, which astronomers discovered in 1998 by examining 10 billion year old Type 1a supernovae scattered across the cosmos, one would have to consider two separate things, the Hubble "constant" and the speed of light. If light really was going much faster in the early universe, we would expect to see an enhanced wavelength redshift in the radiation source light as it reached earth from the very distant and cosmically aged supernovae, as compared to what would be expected from same energy photons released from the same radiation source with today's value of the speed of light. The other parameter to be concerned about is the Hubble "constant", which could itself, as its name implies wrongly, change with space or time. Its definition falls out of General Relativity, and represents the "constant" spatially linear expansion of the universe; but nobody really knows how it might really behave with time (e.g. if the expansion rate of the universe was changing itself over the course of the history of the universe). It doesn't help matters when its measured value is still the source of constant debate, because of the natural wide error associated with its measurement. The latest result we have is from the WMAP survey in 2003, which obtained a result of 72 kilometers per second per megaparsec (a megaparsec is $3 * 10^{22}$ meters of space) +/- 10 percent. Well, 10 percent is an enormous error when you are trying to figure out what its value was over 10^{26} meters of space and 14 billion years. In our earlier discussion of space and time, we discussed the paper by Kolb, Matarrese, Notari, and Riotto, where it was suggested that a local observer could be fooled into believing the universe was accelerating its expansion by assuming only a spatially linear Hubble constant expansion of the entire cosmos. For instance, a larger Hubble constant in the outer cosmos, compared to that we have measured in the local universe, would lead to a greater-than-expected recessional velocity of a cosmically distant source as measured from earth. However, the paper doesn't try to cover other cosmological problems with our local universe, except to assume the existence of dark matter; and of course, it assumes that inflationary theory would have caused the perturbations with space.

So then we come back to the constancy (or lack thereof) of the speed of light, and what to make of all the constant measurements, circumstantial evidence of change, and some of the seeming inconsistencies in them. One thing we observe right away is that the results which are obtained in the same way (except for the very early astronomical aberration results), and with higher resolution (i.e. the use of atomic clocks), give very consistent "constant" values for the speed of light. The downside here of course is that measurements done with atomic clocks have only been done for about 40 years now. It also seems clear that gravitational influences may have played a role in early measurements done in previous centuries; earth was rotating slightly faster due to a slightly different mix of tidal forces in our solar system; and we know more now about our orbital pattern around the sun. Tests of relativity theory regarding the speed of light in different inertial reference frames have held up perfectly, as Einstein's equations required. For instance, decaying particles that are accelerated near the speed of light at CERN or Fermilab show exactly the expected extended lifetimes that relativity predicts for a slowing clock in their relativistic reference frame. We also measure only one consistent speed of light coming from different light sources at different locations throughout the local universe, indicating that the speed of light seems to be independent of source recessional speed and time. So the issue is finally settled, right? Well not quite.

In 1920, after Albert Einstein had rewritten his Special Theory of Relativity to include the more General Theory which incorporated gravity, he wrote that there was no reason to expect that the velocity of light would remain the same under all inertial circumstances where gravity wasn't taken into account, as assumed in the Special Theory. What he meant by that, to the surprise of all those who had been indoctrinated with the theory of the constancy of light speed, is that if light (according to the General Theory) can curve around an uniformly warped spacetime field due to the gravity of some massive object in the way (e.g. a star), then its velocity could also be dependent on its position in the field. Whoa! What a revelation at the time for physicists! And of course, Arthur Eddington, when he made that discovery in 1919, he must have been wondering what was really going on here. Well, herein lay the loophole, if you will, in the speed of light story. It won't explain any biblical fundamentalist interpretations of the age of the universe, but it could go a long way to explaining some very fundamental cosmological problems we still have today because of the general acceptance of the constancy of the speed of light.

First of all, you might be inclined to say that Einstein was really referring to the directional component of the velocity of light, rather than its speed component; but his writings were in the context of comparing the Special Theory with the General Theory so that he really would have been talking about the speed component, and not just direction, especially since it was already known to him by then that light was changing direction, if ever so slightly, around large massive objects like the sun. Thus, by jumping from one reference frame to another where gravity was changing, the speed of light would in fact change also because of the havoc that gravity creates on the rulers and clocks in these different frames. Also, Einstein reluctantly proposed that one would have to go faster than the speed of light in order to pass through one of his wormholes, which we discussed earlier with the Einstein-Rosen bridge; but *anti-gravity* would be required inside the bridge in order to keep it open. Thus, a person would for all intense purposes have to be transformed into an imaginary matrix of particles called tachyons. So clearly, the General Theory did accommodate that light speed could change under gravitational influences and that anti-gravity would play havoc with light or anyone trapped by it; of course the side effects (such as instability and collapse) of wormhole travel would likely be unpleasant to the traveler, but no matter. Where he ran into trouble in believing his own equations however was with the argument of causality. If one went faster than the speed of light, then one could go back in time via one of these wormholes. Thus, in principle, one could go back to a point before one was born, or a spaceship could return before it had left, and that sort of thing. Minor problems really, unless chronology is your choice of poison!

We can give an example of what could happen in natural circumstances under Einstein's theory, by using the fabric of spacetime. We know for instance that light does bend around stars like the sun (or even earth for that matter). If your reference frame went along the path of light, you might have said that no change in the speed of light occurred. But if you were clocking light as it traveled from mars to earth when the orbits were such that one was behind the sun with respect to the other at the time of your measurement, and your reference frame was a straight line through the sun from mars to earth, you would have said that light must have slowed down along the way; that's because these two references frames were not inertial ones, since the sun's pull of gravity was affecting your reference frame in a different way. Now let's play the same game with time. If you drew a straight line reference frame axis from mars to earth when no sun or massive system stood in the way between the two, you would have measured light speed to be very much like that which is measured in a vacuum in a laboratory today on earth, assuming both measurement clocks were always on earth. But now, let's suppose by some quirk in the fabric of spacetime that time itself had briefly become nonlinearly bent between mars and the earth. If light were measured going through this time warp, it would appear and be measured to have a much faster speed from mars to earth than what we would have normally expected, because of the effect that the time warp would have on the time axis going from mars to earth, as measured by a clock on either planet. Today, when we measure light in a vacuum, we measure it in a sufficiently small region such that any warping in the fabric of spacetime would never play a role in the measurement of light speed. Yet such a role could not be dismissed when peering out into the distant cosmos to the edge of the visible universe, or beyond, because of the very distinct possibility of even a severe warping in the spacetime continuum at the edge of the universe, the geometry of which we can never really hope to know beyond doubt because of its likely much farther distance beyond the visible horizon that can be seen from earth.

But now, as for light speed's potential dependence on cosmic time, we could find a solution to today's cosmology problems that preserves the essence of Einstein's General Theory of Relativity by keeping in mind the theory's flexibility in dealing with the fabric of spacetime. A possibility exists within the General Theory that, not only space, but time as well could exist along a curved axis. Einstein had preferred to believe that only space could be curved, but he was shown by a colleague that his equations allowed time to curve as well. The curvature of space was a measure of the expansion or contraction quotient of the universe; but time was supposed to be linear, Einstein thought. That time could curve along the space fabric of our universe would cause a discrepancy in its measurement along any particular spatial coordinate axis of the cosmos, where time was assumed to be linear and uniform across that axis. Not only that, it could start to go backwards as you continue along the axis! Seen in this way, one realizes the cosmos at some outer point in space could become its own natural time machine, as discussed earlier with wormholes and the universe horizon; and it is allowed as a possible solution in Einstein's General Theory! Thus if time does this sort of thing at some point in space, that would say something about light speed at the place where time is slowing down, or indeed stopping; hence, it could even be infinite in principle at this point! That light could have a higher speed at the universe horizon (and at this point it still would have it!), where time hasn't budged beyond 0, would suggest that its speed could still be at the same rate as the interaction rate of a unified force of nature that would still be in place before any spontaneous symmetry breaking took place after the big bang! It has been speculated here in some of our models that this speed could have been 10 orders of magnitude above today's speed of light c, corresponding to the proposed gravitational interaction rate of the universe. Chronology would not break down in this model of a warped space and time universe because one would have to trade space for time and vice versa, such that events in the past which you could reach would have to be elsewhere; and chronology effects would only become apparent near the universe horizon anyway, since the time axis gradient near the universe horizon at time 0 would be quite high and then smooth out quickly away from the horizon. In this case, the speed of light as a function of cosmic time would have to drop off considerably from its maximum point at the start of creation to an almost flat (e.g. constant) value with time as you begin to move away from the horizon to our present value of c. This could certainly explain why light speed shows no noticeable effect for us here in the local universe, far away from the universe horizon, and yet have changed just a little bit far out in the cosmos (as suggested by the distant quasar fine structure data); but it could have rapidly changed during and very soon after creation itself, such as during the inflationary period of the universe, and in fact be a possible explanation for inflation which people try to describe in terms of today's timepieces! Having likely occurred far beyond the visible horizon seen from earth, it would be very hard to measure this curvature in light speed's dependence on cosmic time at or near the universe horizon itself.

The problem to see this speed change is that any light coming from the distant past would either never have had enough time to make it back to earth yet, or its photonic energy would have redshifted all the way to 0 such that we can't measure it or even see it at all. This is because the speed of light and age of the universe define the most distant point in the cosmos that we can see today from earth, where light had enough energy to escape its source due to the source's recessional velocity from earth. (A Doppler shift causes light to lose energy as its source moves away from earth, resulting in the redshifting of its wavelength as seen by earth.) That's why, for instance, the most distant quasars and supernovae we detect in the sky today are moving away from us at about .9c; presumably we could see objects moving up to speed c away from earth with perfect telescope resolution. Beyond that point, if we could see it, objects would be moving away from us at or above c, if, and only if, the local source's light speed were larger than c because of the changing fabric of spacetime due to the gravitationally spacetime-warped universe near the horizon; and these objects, along with light, would have a velocity approaching the velocity of the horizon itself as a function of their proximity to it As for light, we would never notice it moving above the value c unless or until it crossed space where the time axis showed significant curvature; and light speed would have a larger speed as it approached the universe horizon, at which point it would have its maximum value, corresponding to what it was at time 0, the time of the big bang itself. As we've mentioned, if the forces of nature were unified at this time, as physicists suspect, then that speed would correspond to the interaction rate of the unified force of nature, speculated here to be equal to the gravitational interaction rate itself, proposed earlier as $10^{10} * c$, due to the unified equality of all force field quanta at time 0. It's important to remember here that the principle of the constancy of light speed under constant gravity, as required by the General Theory of Relativity, would be upheld since the warped fabric of spacetime at and near the horizon, due to sharply changing gravitational fluxes there, would be causing the changes in the value of the speed of light (with respect to earth) to match that of gravity. Thus, the speed of light would only seem like it is changing with cosmic time as far as an observer on earth would be concerned, because on earth we could only notice the effect as having occurred in the far distant past, given how long it would take the light in question to reach earth. The closer we could see to the universe horizon, the larger the gravitational flux that might be noticed in the form of an increasing value in the speed of light due to the resulting warping that would be occurring in spacetime. And now you can see why models exist, as we have even discussed here, that speculate on a speed of light as high as infinity at the universe horizon, because only there would time be 0, such that any light moving through a place where time was stopped would necessarily have to have a speed value of infinity, by definition!

One way to look at this problem is to isolate and observe the universe horizon as it constantly expands, and indeed possibly accelerates; yet time doesn't move, or by very little. That means the horizon speed could in fact be infinity, or some high value; whatever it is, the speed of light would have to match it. And at this point along the universe horizon, no spontaneous symmetry breaking of forces would have yet occurred; thus the horizon speed would match the interaction speed of the still single unified force of the universe. One of the reasons we have trouble visualizing some of these things (and there are no graphs in case you hadn't noticed!) is that physicists have often made a common mistake over the years of trying to geometrize space and time pictorially in General Relativity according to some fixed coordinate axis. This is a crucial mistake, as the quantum gravity theorists will tell us. General Relativity allows many fixed solutions to the fabric of space and time such that the only way to really understand it is within the context of relationships. Everything is relative as they say; sorry about that! Time can be thought to be relative to space, and space relative to time, as well as both being relative to themselves as measured by each other; but if you try to fix one with respect to the other, you're probably headed for trouble. Hence, it becomes possible to consider an expanding space with zero time that makes no sense on a fixed coordinate grid; but if you try to make it relative to something else that is measurable, you might make some sense out of it, and so forth. Those who are formulating theories of quantum gravity today, where they actually try to quantize space and time down to the Planck length scale, have only begun to realize this when trying to incorporate General Relativity into their theories.

Now it may be a "bummer", as we used to say in my college days, that we can't see any of this; and you might be tempted to think at this point, at least for a moment, that we're talking about a bunch of Podunk. Well, I did warn you that physicists have funny thoughts at times! But there are some things we can measure and are observed in our present visible universe that give some credence to these ideas. For one thing, we have tried to be consistent with the General Theory of Relativity, which has had remarkable experimental success. For instance, we can measure and predict exactly the extended lifetimes of decaying particles traveling near the speed of light. We can see light being warped by massive objects in space. We can measure the GPS clock signal on a satellite traveling at high speeds in orbit around the earth and see that its clock has actually slowed by the predicted amount. And we can also predict the amount the satellite's clock has increased by being in a smaller gravity well far above the earth, where there is less gravity, and get the right result. Thus, we know there is a time dilation effect due to objects going near the speed of light; and we know objects, including light itself, are affected by the warping of spacetime due to changes in gravity.

Now, it has long been theorized, especially by inflation theory proponents, that some kind of antigravity field gave rise to an expanding universe that has kept the universe from not collapsing back in on itself. But data taken in 1998 from the most distant type 1a supernovae in the cosmos that have ever been seen gave cosmologists more than they bargained for; not only was an expansion of the cosmos confirmed, but it was a surprising *accelerating* expansion of the outer universe that no one had expected at the time. A larger speed of light in the early universe would provide a possible solution to this observation. Yet some cosmologists have quickly tried to explain it under the general terminology of dark energy, while no one really knows what that is. And some have borrowed Einstein's cosmological constant "blunder" from General Relativity in order to explain it, a constant that Einstein only used to try to describe a static universe that he wanted to have, but did not get; however, a "dark energy" constant would have to be no less than 10^{120} to explain the accelerating universe data! And still others, as we have seen, have theorized inflationary Hubble perturbations in the fabric of spacetime are responsible for what is seemingly observed by earth to be accelerating away from us. Finally, there are the fine structure constant measurements reported in 2005 from distant quasars which are ever so slightly different from what is measured today with local source spectra on earth; this could also be explained by a cosmically aged different speed of light value. This splitting, described in quantum mechanics as dependent only on the universal constants electronic charge, Planck's constant, and the speed of light, could only change if somehow at least one of these three constants was affected by time, or something having transpired over the time that the light spectra (from the splitting) traveled to earth over the wide swath of the cosmos, such as changing gravitational conditions in the fabric of spacetime causing a resulting effect on our clocks and rulers. While it would seemingly occur on a much smaller scale away from the universe horizon, that does not mean that some distortion or curvature in the fabric of spacetime could still not be present to some degree in our "flat" and "small" part of visible space, which we have speculated before as likely being a small subset of the entire universe anyway. In any case, only variable speed of light theories can explain all these issues at once; and General Relativity makes them possible if changing gravity or anti-gravity conditions in the fabric of spacetime are involved.

Thus, there is the observed evidence of the curvature of light around the sun and other massive star systems, due to the warping of spacetime by their gravitational fields. It shows that light does indeed depend on gravity as it is supposed under General Relativity; one could guess then what its imaginary behavior would be under the influence of anti-gravity, although we haven't been able to produce anti-gravity yet. The Casimir effect is about as close as we can come to producing anti-gravity today; but it only can be seen on a miniature scale anyway. Perhaps it's more understandable to say that light (and light speed) depends on any warping in spacetime, which gravity or anti-gravity would cause. Theoretical Einstein-Rosen bridges could also be a natural consequence of any warping in the continuum of spacetime, allowing for the possibility of wormhole creation in a supposed anti-gravity environment, where one could cross over the fabric of spacetime in a hurry; but natural wormholes have not been found, and the creation of one could lead to chronology difficulties.

Where this is all leading us is that universal warping in the fabric of spacetime could explain all of today's cosmology problems, both local and distant. A seeming change in the speed of light where warping of time may have occurred in the outer visible cosmos, due to presumably changing gravitational effects, could be the cause of the observed accelerated expansion of the outer cosmos. Distant objects moving away from us, for instance, would show a seemingly increased acceleration, where any time warp would not be accounted for by the Hubble constant. When distant expansion rates are measured, one measurement is done of distance (l), by measuring the luminosity of the source, and another measurement is done of the source's light wavelength Doppler red shift, in order to obtain its recessional velocity (v) from earth. The ratio of v to l gives us the Hubble constant, according to General Relativity. Hence any warping in the fabric of cosmic time that is crossed by the source light would not be accounted for in this simple framework, especially the further out we go, and thus could result in a seeming increase in the speed of light from the distant past, as measured by a clock on earth; in reality, assuming the relativistic Doppler shift of the source object's speed has been accounted for, any further increase in light speed from the distant past due to time warping (from a constant energy source) would enhance the wavelength shift as the light reaches earth. Hence, it would be a speculation that light received by earth is much more greatly Doppler shifted from radiation sources that are nearer the universe horizon, where this time warping in the fabric of spacetime could occur in earnest. The normal Doppler effect, due to a source's recessional velocity, and the Hubble constant itself wouldn't be able to account for this warping because time is usually assumed to be linear in most calculations with General Relativity, including those preferred by Einstein; as a result, it could explain why greater than expected red-shifted wavelengths coming from distant type 1a supernovae are what are observed. The actual wavelength shift in the 1998 supernovae data suggested the observed source light was about 10 to 15 percent farther out than expected. One can only imagine how much more the data might be shifted if it had originated from near the universe horizon itself where universal spacetime geometry would be at its extreme curvature; the data could be Doppler shifted all the way down to 0, were it coming from this region, because the universe expansion horizon could well be going at the speed of its local speed of light, which could itself be much above c!

Problems such as the Pioneer anomaly and galaxy rotation rates that are too fast (given what we know about the visible matter seen in those galaxies) could be explained by an ongoing warped geometric or gravitational expansion of space. Measurements of data points in these cases are done within a close proximity of time by comparing different coordinates of an object of observation that are within a close proximity of space. Thus any warping that is already present in the fabric of spacetime would not explain these effects like it could with the supernovae data, because the phenomenon would effectively be canceled by the subtraction of one measurement from the other; but any warping of spacetime *during* the interval of traversal of an object between two measurement points could be seen by the observer in some fixed reference frame. If we assume time behaves linearly in local space where the measurements are made, we can speculate then about why spatial warping could be occurring dynamically within the time of the traversal. The expansion of space that is described and predicted by General Relativity is not enough to explain these issues, since the recessional velocities of the objects in question are not anywhere close to the speed of light that would be necessary to justify an unexplained inner observed anomalous acceleration of cH_0 (H_0 being the Hubble constant); this anomalous acceleration appears to be the same for *both* the Pioneer anomaly and the galaxy extraordinary rotation rates. But if the universe were closed in some way, or wrapped in on itself in some way (for instance by a string theory dimensional description of the universe or a gravitational spatial warping of the entire universe not unlike we observe with massive stars), such that the universe was effectively a closed hyper geometrical sphere, the trajectory of light itself in the vacuum of space could in principle take a warped circular path along all points in space with inner acceleration (directed toward the center) equal to cH_0 and radius of c/H_0, just as light is observed to do when traversing the gravitational field of a massive star. If so, the radius of curvature of light traveling today in pure space would equal the distance that light could have traveled over the history of the universe at speed c, or 10^{26} meters. This would mean for instance that light, at the circumference of the visible universe horizon would actually be traveling in a circle; if time were stopped at the visible horizon (as predicted by General Relativity for anything traveling with the speed of light), then an observer looking out along the horizon would simply see the back of himself or herself, as if you were looking at the back of your head through mirrors positioned in front and in back of you! At other points in space, you might be able to see the back of your head also; but with time ticking as it does in a fixed reference frame and the speed of light being as it is, you would be waiting for a while, probably a long while in fact, as much as 2 * pi * the visible radius of the universe divided by c, or about 85 billion years or so! However, each ray of light would have its own focal point in three dimensional space such that the universe continues to look flat and uniform to a fixed observer.

Another bizarre possibility could be that all entities of quantum space wrap around in a circle in hyper dimensional space with some hyper dimensional radius that is defined by a *common* focal point in hyper dimensional space, such that all objects and points continue to have a uniform inner acceleration in three dimensional space of cH_0. Whatever the reason, the Pioneer anomaly and faster than expected galaxy rotation rates both have this same inner acceleration discrepancy directed inward toward the center of orbit; in the case of faster than expected galaxy rotation rates, it is an object of observation in orbit about its galaxy, while in the case of the Pioneer anomaly, it is the point of observation's orbit (the earth) about the sun! Whatever the theory, there does seem to be something out there that has a relative receding velocity of c traveling along a circular path connecting an object with its point of observation, should a parallel with the General Theory of Relativity ever be made. For those who snicker at these ideas, you might need to hold some back for those solutions which involve dark matter, because it will never become a solution to both problems. Conventional cosmologists would rather have us all forget about the Pioneer anomaly, because they know that the still popular theory of dark matter would have to become asymptotically impossibly huge as we go farther out into space in order to explain the effect so close to home; and distant galaxies are just not rotating that fast!

So finally, we come to the theological point of this discussion; and implications of this analysis are enormous indeed. For one thing, we have to challenge any idea of a 6000 year old universe, even with a much higher speed of light in the early universe; and the speed of light by itself would not be dependent on cosmic time, in agreement with relativity theory, but rather dependent on the universal fabric of spacetime that may well be changing over time with respect to the continual expansion of the universe and changing gravitational flux in the universe. And at the universe horizon, this flux could have a sharp gradient and be especially intense such that sharp local changes in the speed of light could have occurred, and thus mean that light speed in the very early universe could have been much greater than we observe today on earth. Whatever the case, the key to the universe and the Theory of Everything lies there; if we could see the universe horizon we would have a snapshot of creation in the making at time 0 itself, and thus find the unification of all forces of nature in the process. This could mean then that the universe is wrapped in on itself in some hyper dimensional way, such as explained by string theory; how else would the farthest point in the cosmos (i.e. the universe horizon) be also its starting point (time 0)? Thus, in some geometrical sense, the hyper universe could have an overall symmetry that God prefers. But one thing we do know about the preference of God is that he seems to have a thing for circles and spheres; and no place is that more apparent than with the baffling problem of the Pioneer anomaly where the only physical explanation presented by those that are studying the phenomenon is that every point in space seems to act like the center of a sphere that has an inner acceleration cH_0 directed toward the center point at all times. Thus, if God enjoys geometrical symmetry, as he seems to do with the path of light about the visible universe, symmetry's most important feature will always be its center, or its focal point. In other words, if God were to be found at the center of it all, he seems to crave attention!

XV. Conclusion

Richard Feynman used to say that science is the belief in the ignorance of the experts. If that is the case, I would think then that religion is belief in the rejection of intelligence of everyone else. The Bible for instance points out that the foolishness of God is greater than the wisdom of man. When we consider the awe and majesty of his universe, even with all of its loss of symmetry, imperfections, violations, finitude, and final doom, it sure got that right. When we take into account the arrogance of man, especially of those who sometimes are even scientists, we can get all sorts of bad and biased results. Theologians and people of faith can easily be found just as guilty; when they say they're right because everyone who disagrees with them is wrong, then they too have a problem. The trick for humanity is how to get people of different backgrounds, skill sets, interests, and beliefs to come into agreement on anything. With all the different races, creeds, beliefs, religions, sects, and tribes running around, it's truly a wonder of nature that we haven't destroyed ourselves by now! It clearly seems obvious that God wanted to have this diversity; or to invoke the Anthropic Principle, it exists this way because nature required that it exist this way in order for us to exist this way. One of the ways that God makes it work, for instance, is through the shared consciousness of the world; there is a voice that is present in all of us which seems to come from a common source. That common source speaks to us through some physical interface in our conscience that allows us, in the absence of personal greed, envy, pride, or evil, to think in unison, and hence gives us a window into the mind of God.

When we talk about our image of God, we all have our personal tastes and views on the matter. Albert Einstein, for all of his references to God, once finally stated "I do not believe in a personal God" and goes on to say, "If something is in me which can be called religious, then it is the unbounded admiration for the structure of the world so far as our science can reveal it." Now experience has shown that scientists who disagree with Albert usually end up doing so at their own peril. So, far be it for us to try to take on one of the greatest scientists of all time, even if it involves a field outside his jurisdiction, namely God. However, one can put his comments into some perspective. With all of our appreciation for the majesty of the universe, and our ability to understand it, it is a hard thing to imagine that we humans could have such a thing as a "personal" God, as if he was somebody's personal pet. But by the same token, it is also a hard thing to imagine that we humans live in such a majestic universe without somebody having had some plan of action, an intelligent design if you prefer. And Einstein's frequent and famous references to God certainly suggest that he had wondered seriously about it. That design gave structure to the universe; but Einstein desperately wanted science alone to reveal it.

That science could suggest intelligent design from the Second Law of Thermodynamics itself is a bit ironic; it suggests that a closed system will not only be imperfect, and certainly not symmetric, but must in fact become chaotic with time, which of course the universe is not, as it continues its orderly expansion; thus if the universe is part of even a grander symmetric design of the hyper universe, then a closed system is implied, giving rise to a necessary order requiring some *intelligent* design somewhere. And the fact that science hasn't been able to describe the total structure of the universe speaks volumes about the pieces that are likely to be missing in our miniscule world of space and time. If there are other dimensions out there that we simply can't see, then there is probably an ultimate symmetry behind it that would explain everything. The case that an ultimate symmetry exists somewhere behind our universe can be made, for instance, by the fact that everything that has ever occurred or ever existed in our universe resulted from some form of symmetry breaking; and if there is symmetry, then there is a center to which and from which everything is relative. That center provides the energy for all existence; and people of faith will refer to this source as being God. As God seems to crave attention, who would we be to disagree! Likened to string theory's wrapping of the hyper dimensional universe about its center, the ultimate symmetry of the universe would be revealed in the wrapping of this center about the rest of the universe out to its horizon, which hence would hold the key to our understanding of the Theory of Everything.

Burkhard Heim may in fact have gotten as close as anybody to finding it; it's just amazing what the blind see sometimes! For fifty years he worked on a multidimensional theory, starting from General Relativity, to try to describe all forces of nature in the same way, with a model that is quite analogous to string theory today. After he died in 2001, his colleagues continued his work to predict particle ground and excited energy states. The observed agreement with his theory has been astounding, out to 7 decimal places! His mass formula has predicted elementary particle masses down to one part in 1000; and the errors that do exist are easily explained by our knowledge and resolution of Newton's gravitational constant, which is used in the calculation. In many ways Heim's theory has gotten closer to finding the Theory of Everything than the standard model or quantum gravity theories (which include string theory) because it has accurately predicted masses, particle excited states and particle decay times that the other theories have so far been unable to do, even today. The fact that he has not received recognition in the physics community is quite interesting, being that he avoided academia due to his disabilities in life; and it didn't help that he had a distinct distrust for technical journals. He found them useless because he thought people would only plagiarize his work; the downside of course was that not only were people less aware of his work and his difficult mathematical notations, but also no one could be sure of their authenticity, since very few theorists have had the opportunity to rigorously defend or challenge the work. The fact that Heim is typically ignored by the greater physics community is therefore unfortunate. If indeed mass itself could be explained by projections from other dimensions, then spiritual existence itself could be explained this way; and hence one could even model, for instance, how Jesus with a resurrected spirit could have conversed with Mary and his disciples, yet have been able to walk through doors and digest food!

Fortunately, all scientists haven't dismissed Heim theory. Some have even tried to use it as a basis for quantum gravity theories, such as string theory. And in July, 2005, the Nuclear and Future Flight Propulsion Technical Committee of the American Institute of Aeronautics and Astronautics (AIAA) gave its "best paper of the year" award to articles published by former colleagues of Heim, W. Droescher and J. Haeuser (***Guidelines for a Space Propulsion Device Based on Heim's Quantum Theory***, AIAA, 2004-3700***; Heim Quantum Theory for Space Propulsion Physics***, AIP Conference Proceedings, V746, pp.1430-1440, February 6, 2005); in these articles, they discuss Heim theory applications for future aerospace technology. It should not surprise anyone that Heim might have had the right idea about mass. The fact that space and time are warped by mass to create gravity, and that even quantum mechanics shows an inverse relationship between a particle's mass and its effective size, should have told us something. Later in his life, Heim himself was interested in including consciousness and the spirit within multidimensional layers of his theory; and those developing models of string theory today are also trying to do some of that, given their likeness to existing models of consciousness that some psychologists use. God may not need to be a Trinitarian; but he sure does seem to show interest in multidimensional math and algebra!

And when we talk about the multidimensional aspects of string theory or Heim theory, we come to the mystery surrounding the Pioneer anomaly and how every point in space in our expansive universe seems to form a perfect sphere about itself with a constant inner acceleration cH_0, as observed by the Anderson team, when they conclusively identified the Pioneer anomaly as a real effect and without explanation. The wrapping of hyper dimensional space about our own could provide such an explanation, for instance. An illustration of this idea can even be provided by General Relativity, which only assumes our own dimensions of space and time; it suggests that something traveling with recessional velocity v (with respect to earth) along the circumference of a circle with radius r has an inner acceleration (directed toward the center) of $vH_0 = v^2/r$, where H_0 is the Hubble constant. In the case with the Pioneer anomaly and observed galaxy rotation rates, the anomalous inner acceleration of the observed objects would be explainable by something traveling between the object points of measurement in a circular path with a velocity of the speed of light itself ($v=c$), as if there were a warping of the path of cosmic photons due to the continually shifting gravitational geometry of the expanding universe.

The enlightened folk of the 18th century used to believe that angels pushed the planets around the sun with their wings. This tidbit must have amused Richard Feynman, because he pointed out that maybe they were not so wrong after all; they just had them pushing in the wrong direction!   With effects like the Pioneer anomaly that even Feynman didn't know about, perhaps those angels had even another surprise in store for us! Clearly linked to the fabric of space itself, something does seem to be pushing not just the planets around the sun but also all space about itself, even as the outer cosmos expands and accelerates outward. Some physicists refer to a dark matter and dark energy scenario to explain this mystery. But perhaps the universe isn't so dark after all; for instance, dark matter will never be able to explain both the Pioneer anomaly and fast galaxy rotation rates at the same time; but a warped trajectory of light in geometric space about a circle of radius c/H_0 could explain both. And if the warped fabric of space and time were the cause of a changing in light speed in the early universe, as measured by earth, one could also explain an expanding outer universe that appears to be accelerating without a need for dark energy. Those angels would be allowing the nostrils of God to blow inward on the entire universe, even as he was stretching the heavens apart. In any case, the observed effect seems to be working. The universe has enough density to maintain cohesion and not rip itself apart (at least not yet) into a Big Chill, while at the same time it doesn't have too much density to simply collapse back on itself into a Big Crunch. Perhaps that doesn't make you feel any better; but God seems to know what he is doing.

Now what makes this story even more compelling, for us earthlings anyway, is a bizarre observation that if the angels had replaced c with g, where c is today's speed of light and g is the speculated primordial universal gravitational interaction rate whose value has been guessed here in earlier discussions to be 10 orders of magnitude above c, they would have obtained an inner acceleration, gH_0, for the entire universe about some common point; it turns out this value happens to be very close to that of today's gravitational acceleration on earth itself (9.8 meters/second^2)! Perhaps the geocentrists of the 16th century and the enlightened folks of the 18th century weren't so whitewashed after all, although things may have turned out in a much different way from anything they would have guessed! Could there thus be something special about our mass to size planet ratio on earth that attracts the right elements and minerals for life sustenance, according to the Anthropic Principle for instance? And then there is the matter of our seemingly unique life preserving stable stellar orbit in the heavens. Whether or not there really is something special about earth among the cosmic oasis of planets, God does seem show an interest in our planet; and whether or not we are really just playing with numbers, the values of cH_0 and gH_0 certainly seem to show that God has a thing or two in our universe for speed, symmetry, cohesion, and coordination!

St. Paul, who didn't know about any of these things, used to say that God is through all, in all, and part of all things in the universe; if he had been a physicist, what he really might have tried to say was that God is the focal point of a perfect realm which provides symmetry to the whole universe through which all things must pass. And this was a point that Einstein desperately tried to avoid and reject when he considered the concept of God; yet he could never evade it. For him, it was disturbing that the expanding universe could be mapped backward in time to a common starting point; yet that is exactly what his equations were telling him. When he imposed his cosmological constant on his theory to make the universe static, he created his own worst "blunder". It was the theologians, to his dismay, who were found to be in total agreement with him!

But physicists today, as Einstein was before them, are seriously pursuing the Theory of Everything, and one which will explain mass, our existence, the unification of all forces of nature (at least as it must have existed in the very early universe), and indeed the missing symmetry of the universe itself. The laws of physics will eventually help us to find the formula; but not unlike General Relativity Theory, if we never really understand how to use them, we'll hopelessly be lost how to apply them. And then Valentinus would have been right all along; the human fall from grace is characteristic of our place, and ultimate demise, in the universe. If we don't find our way out of this trap, God will certainly condemn us, if for no other reason, we made no effort to try to get out. And perhaps, that in turn is what our relationship with God is all about. His status may transcend our universe in a way that mirrors the symmetric dimensions of the hyper universe, both too small and too large for us to see, and all at the same time! The ultimate symmetry of the universe must be its generator, creator, perfecter, and master; without it, symmetry is broken into spontaneous disruption, distortion, and finitude. Unfortunately, that is the state of the present snapshot of the universe in which we live; but we really shouldn't be so surprised. Until we find peace from war, love from hate, trust from suspicion, and truth from falsehood, we're never really going to get there, back to a state and time in which our universe was unified, united, perfected, and symmetric; yet it will only be then that we truly find peace, and our own symmetry, with God.

When considering quantum mechanics, Richard Feynman, never being one short for making a quotation or two, said that in the final analysis no one really understands it. Of course, someone must have thought he did, as he was picking up his physics Nobel Prize in 1965 for his breakthrough work on quantum electrodynamics (one of my favorite physics subjects, as it turns out, probably because I understood it better than everything else). The problem is of course that we can only see things properly in a four dimensional space and time matrix because that is the limited range of our physical detector.

If we could just see out beyond those dimensions into another outer realm, one would perhaps be able to ascribe exactness to quantum mechanics, for instance, in the supposed hyper universe, which then could provide an exact description to our universe itself. However, God seems to have made this idea impossible, by limiting us to those four dimensions of space and time. Particle physicists help us resolve some of these issues by giving us rules about how certain interactions must behave; but they also create more problems along the way, such as has happened with the cherished CPT theorem, if quantum field theory is to be believed. But as the CPT problem seems to suggest, there probably is a quantum number missing, perhaps inside another dimension, which could give back quantum field theory its lost symmetry, and maybe even give hope for quantum mechanics to become more than just a theory of probabilities. If so, God might not have been playing dice with the universe after all; and Einstein would have been relieved!

And let's not forget about General Relativity. Lots of people try to describe it; but lots of people get lost trying to understand it. And part of the problem is that we still get caught up trying to think inside the box; and yes, physicists too aren't immune to doing that! When we start to peer outside that box, General Relativity opens up many possibilities for relational analyses within just the fabric of spacetime itself. For instance, funny things can happen on a relative coordinate grid that you wouldn't expect to have happen on a fixed coordinate grid. You could have light traveling at infinite speeds along a universe horizon where time is at a standstill; but space is still expanding. And as the spacetime curvature of the universe flattens out, you would get a constant limited value of the speed of light that we see today in our visible end of the universe. But it would also turn out that light, in a still unified force a nature, along with gravity, traveling at a rate of 10^{10} orders of magnitude above c, would have easily traveled the full range of the visible universe we see today (10^{26} meters) in just one year, if the forces had maintained unification! Of course, they didn't; but that doesn't mean gravity couldn't still have had that large an interaction rate. That certainly fits nicely with models of how the universe got to be as big as it is in the first place (e.g. inflation), but it doesn't help us in trying to narrow down the age of the universe to something small, such as 6000 years of recorded biblical human history. And that's because, among other things, light isn't the end of the story. Many of the elements that make up the cosmos would have taken much, much longer to form; and the linear expansion rates of various points in the cosmos itself (the sources of whose light are traveling at today's known light speed) suggest a much longer universe, measured to be 13.7 billion years of age. But if it makes anyone feel any better, the last mutation of the human brain may have occurred about 6000 years ago, which could explain why some things are only written down as they are in the Bible.

And then there are those wormholes, theorized in science as nonlinear instabilities in the fabric of spacetime, and interesting because they give promise to the idea of transgalaxy or transuniverse travel, or even time travel. While no such instability has been shown to exist in nature, there are absurd ways with enough energy in a small enough instability of space and time to create one theoretically, a mini big bang if you will. But the risk associated with going through such a thing would quite likely outweigh the benefits, even for a desperate and cosmically advanced civilization trying to escape the impending finitude of our life-sustaining universe. If it were attempted however, the hyper dimensions one might be cast into could tell us something about the lost symmetry of the universe, or the realm of God for that matter. And as for natural wormholes, were they to be found, the possibility would exist for a possible shortcut to be created across space and time, thereby creating all kinds of chronology problems, but not likely to occur because space and time would necessarily become disjointed and exchanged such that paradoxical causal events should not occur.

When I finished graduate school, or more specifically, on the same day that I passed my thesis defense, the university, for reasons never fully understood by me, but yet purely coincidental, decided to move a big rock in the center of campus from its original place to another place about 10 meters away. "The Rock", for those not familiar with all things Northwestern, is that symbol of university spirit, camaraderie, unity, and goodwill that for one reason or another allows itself to be painted by some university organization almost every day of the year from goodness knows what to goodness knows whatever. Anyway, that day was August 16, 1989; and it was the day the movers accidentally split "The Rock" into two pieces while it was being moved. Of course, it would get put back together quickly with a slab of concrete; but that is an event that will go down into Northwestern folklore and history; and it is the day the university showed me the door to the world. I can spend a lot of time thinking why this happened when it did; but "The Rock" apparently became a symbol for me as well. If it is supposed to represent the university's soul and symmetry, being in the center of campus, then on that day, the concept of spontaneous symmetry breaking took on a whole new meaning. As for me, you could say time undertook a phase transition from one universe into another. As life moves on for all of us however, the wind of God is constantly in motion; and our own rock of life usually symbolizes something much more stable and profound. Thus, the question is: are we universe dwellers ready for that time when we undertake another phase transition, or do we let that rock simply undergo a renewed symmetry breaking?

God may have indeed provided for us a perfect symmetry, since it seems to be in his nature to be at the center of it all, space and time inclusive. As for a name for this idea though, I guess you could call it GUSS theory – Grand Unification by Super Symmetry! The name certainly sounds serious enough! But the final drawback (and convenience) of it is that it could never be proven! And such is the fate for the clash between the cultures of church and science, in all seriousness. However, if there were ever to be any final coordination and cooperation between the two cultures, it would serve them well to adhere the lesson of St. Paul, who wrote in his biblical letter to the Philippians, "Whatever is true, whatever is honorable, whatever is just, whatever is pure, whatever is lovely, whatever is gracious, if there is any excellence, if there is anything worthy of praise, think about these things." And this at least is something that physicists, when true to their profession, will try to do everyday.

Einstein was right in fact; and he didn't even know it at the time: the total mass energy of the universe is counteracted by the total gravity that is in the universe. For every positive charge created, there is a negative charge produced that counteracts it. For every action in space, there is a reaction in the opposite direction, and so forth. The fact that we sometimes don't see the ultimate symmetry in the universe, such as with mass, is maybe because our laboratory in the physical universe is simply not large enough to see the hyper universe that may still be out there; and our detectors are not sensitive enough to pick up everything that has gone missing. If, at the end of the day, one finds that so-called hyper universe, one may find that all things in space and time are wrapped up together working in perfect synchronization and harmony and symmetry with each other, driven from a common center, as many physicists have hoped to find and understand. And people of faith, and even some theologians, will appreciate the fact that this driver of ultimate symmetry is its most fundamental feature, the center of it all, the "I am" of the universe, parallel to the "I am" at the center of our own consciousness, the Yahweh in all of us, whom they call God.

Bibliography

Adey, W.R., *Electromagnetic Fields, the modulation of brain tissue functions – A possible paradigm shift in biology*, International Encyclopedia of Neuroscience, 3rd ed., Elsevier, 2003.

Albrecht, A., Magueijo, J., *A time varying speed of light as a solution to cosmological puzzles*, Phys.Rev. D59, 043516, 1999.

Anderson et al., *Study of the anomalous acceleration of Pioneer 10 and 11*, Phys. Rev. D65, 082004, 2002.

Barger, V., Marfatia, D., and Whisnant, K., *LSND anomaly from CPT violation in four-neutrino models,* Phys. Lett. B 576, 303, 2003.

Bell-Burnell, Jocelyn, *Pliers, pulsars, and extreme physics*, Astronomy and Geophysics, Blackwell Publishing, February 2004.

Bulletin of the Atomic Scientists, *60 Years Ago...*,V58, N06, pp54-55, Nov/Dec 2002.

Christianson, Gale, *In the Presence of the Creator: Isaac Newton and his Times*, The Free Press, 1994.

Chicago Sun-Times, *Human Brain is still evolving, U. of C. researcher finds*, Sept. 9, 2005.

Darwin, Charles, *On the Origin of Species by Means of Natural Selection, or The Preservation of Favoured Races in the Struggle for Life,* 1859.

Davies, P.C.W. and J. Brown, *Superstrings: A Theory of Everything?*, 1988.

Droescher,W., Haeuser,J., *Guidelines for a Space Propulsion Device Based on Heim's Quantum Theory*, American Institute of Aeronautics and Astronautics, 2004-3700.

Droescher,W., Haeuser,J., *Heim Quantum Theory for Space Propulsion Physics*, AIP Conference Proceedings, V746, pp.1430-1440, February 6, 2005.

Ellman,R., *The Origin and Its Meaning*, 2nd ed., The Origin Foundation, 2004.

Feynman, R. and Leighton, R., *"Surely You're Joking, Mr. Feynman!"*, W.W. Norton, 1985.

Feynman, R. and Leighton, R., *What Do You Care What Other People Think?*, W.W. Norton, 1988.

Ford, E.B., Lystad, V., Rasio, F.A., *Evidence for Planet-Planet Scattering in Upsilon Andromedae*, Nature, 14 April, 2005.

Frayn, Michael, *Copenhagen*, Anchor Books, 1998.

Fudjack, J., Dinkelaker, P., *The Enneagram as Classic 'Double Mandala'*, Part I-II, The Enneagram and the MBTI, March-April, 1999.

Gribbon, John, *Companion to the Cosmos*, Weidenfeld and Nicolson, 1996.

Growcott, G.V., *"The History of the Doctrine of the Trinity"*, the Antipas Organization.

Harris, Kevin, *Collected Quotes from Albert Einstein*, 1995.

Hawking, Stephen, *A Brief History of Time*, Bantam Books, 1988.

Hoeller, Stephan, Valentinus: *A Gnostic for All Seasons*, Gnosis,

1985, 24.

Hoagland, R., *Context and Implications of the Discovery of Extraterrestrial Life: A Whitepaper*, 1989.

Hubel, D., *The Brain*, Scientific American, pp.38-47., Sept. 1979.

Knight, R., *What 'gay marriage' Really means for America*, WorldNetDaily.com, May 19, 2004.

Kolb, E., Matarrese, S., Notari, A., and Riotto, A., *Effect of inhomogeneities on the expansion rate of the universe*, PRD 71, 023524, 2005.

Kolb, E. and Turner, M., *The Early Universe*, Addison- Wesley, Menlo Park, Ca., 1990.

Lawrence, Andrew, *God does not play dice with the universe*, Life Purpose Society, 2003.

Layton, B., *The Gnostic Scriptures*, Doubleday, 1987.

Lederman, Leon, *The God Particle: If the Universe is the Answer, What is the Question?*, Delta, 1994.

Levy, et al., *Reassessing the chronology of Biblical Edom, new excavations of 14C dates from Khirbat en-Nahas (Jordan)*, Antiquity 302, December 2004.

Los Angeles Times, *The Scientology Story, Defining the Theology*, June 24-29, 1990.

LSND Collaboration, *Evidence for neutrino oscillation from muon decay at rest*, Phys. Rev. C54 2685-2708, 1996.

Martin, W., *The Kingdom of the Cults*, Bethany House, 2003.

Montgomery, A., Dolphin, L., *Is the Velocity of Light Constant in Time?*, Galilean Electrodynamics, Vol. 4, no. 5, Sept/Oct 1993.

National Institute of Standards and Technology, U.S. Department of Commerce, 2003.

New Catholic Encyclopedia, Vol. X, McGraw-Hill, 1967.

Nobelprize.org, The Nobel Foundation, 2005.

Owen, R., A *Brief SETI Chronology*, The SETI League, 2003.

Pello et al., *ISAAC/VLT observations of a lensed galaxy at z=10.0*, Astronomy and Astrophysics, V416, page L35, 2004.

Penrose, R., *Shadows of the Mind*, Oxford, 1994.

Peterson, Philip, *Cosmic Acceleration: Hounding the White Whale of Cosmology*, Empyrean Quest Publishers, 2004.

Physics Web, *Over the top*, 9 June 2004.

Physics World, *Peter Higgs: The man behind the boson*, July 2004.

Physics World, *Solar neutrino puzzle is solved*, July 2001.

Pope John Paul II, *Address to the Pontifical Academy of Sciences*, 3 October 1981.

Powers, T., *Heisenberg's War: The Secret History of the German Bomb*, Knopf, 1993.

Riess, et al., *Observational Evidence from Supernovae for an Accelerating Universe and a Cosmological Constant*, The Astronomical Journal, V116, p.1009-1038, 1998.

Ritschl, Albrecht, The Christian Doctrine of Justification and Reconciliation, Clark, 1900.

Rosenkrantz, H., and Roberts, W., Bloomberg.com, *Wilson's Iraq Assertions Hold Up Under Fire From Rove Backers,* 14 July, 2005.

Sagan, Carl. *The Dragons of Eden*, Random House, 1977.

Schaefer, Henry F., *Scientists and their Gods*, Leadership U., 1999.

Science Daily*, Scientists Discover One of the Constants of the Universe Might Not Be Constant*, May 12, 2005.

Setterfield, B., Norman, T., *The Atomic Constants, Light, and Time*, Invited Research Report, Lambert Dolphin Research Library, August 1987.

Singh, S., *The Big Bang: The Origin of the Universe*, HarperCollins, 2004.

Super-Kamiokande Collaboration, *Evidence for Oscillation of Atmospheric Neutrinos*, Phys. Rev. Lett. 81, 1562–1567, 1998.

The Holy Bible, Revised Standard Version, Thomas Nelson and Sons, 1952.

The Philosophy Shop, *Albert Einstein: God, Religion and Theology*, Copyleft, 2005.

The Philosophy Shop, *Quantum Mechanics: Richard Feynman*, Copyleft, 2005.

The Sunday Times –Britain, *The Downing Street Memo*, May 1, 2005.

The Universe Today, *Gravity Moves at the Speed of Light*, January 9, 2003.

Thompson, R.F., *The Brain. An Introduction to Neuroscience*, Freeman, 1985.

Troitskii, V.S., *Physical Constants and Evolution of the Universe,* Astrophysics and Space Science, 139, p.389, 1987.

Wakin, Edward, God *and Carl Sagan: Is the Cosmos Big Enough for Both of Them?*, U.S. Catholic, 1981.

Weisstein, Eric, *World of Physics*, 2005.

Westfall, Richard, *The Life of Isaac Newton,* Cambridge University Press, 1993.

Wikipedia, The Free Encyclopedia, 2005.